ISBN 978-1-5281-1577-3
PIBN 10906877

1 MONTH OF
FREE
READING

at
www.ForgottenBooks.com

By purchasing this book you are eligible for one month membership to ForgottenBooks.com, giving you unlimited access to our entire collection of over 1,000,000 titles via our web site and mobile apps.

To claim your free month visit:
www.forgottenbooks.com/free906877

English
Français
Deutsche
Italiano
Español
Português

www.forgottenbooks.com

Mythology Photography **Fiction**
Fishing Christianity **Art** Cooking
Essays Buddhism Freemasonry
Medicine **Biology** Music **Ancient
Egypt** Evolution Carpentry Physics
Dance Geology **Mathematics** Fitness
Shakespeare **Folklore** Yoga Marketing
Confidence Immortality Biographies
Poetry **Psychology** Witchcraft
Electronics Chemistry History **Law**
Accounting **Philosophy** Anthropology
Alchemy Drama Quantum Mechanics
Atheism Sexual Health **Ancient History**
Entrepreneurship Languages Sport
Paleontology Needlework Islam
Metaphysics Investment Archaeology
Parenting Statistics Criminology
Motivational

PEABODY MUSEUM OF NATURAL HISTORY
YALE UNIVERSITY
BULLETIN 27

The Type Fox Hills Formation, Cretaceous (Maestrichtian), South Dakota

Part 1. Stratigraphy and Paleoenvironments

KARL M. WAAGE

Peabody Museum of Natural History and Department of Geology
Yale University

NEW HAVEN, CONNECTICUT
1968

Bulletins published by the Peabody Museum of Natural History, Yale University, are numbered consecutively as independent monographs and appear at irregular intervals. Shorter papers are published at frequent intervals in the Peabody Museum *Postilla* series.

The *Peabody Museum Bulletin* incorporates the *Bulletin of the Bingham Oceanographic Collection,* which ceased independent publication after Vol. 19, Article 2 (1967).

PUBLICATIONS COMMITTEE: A. Lee McAlester, *Chairman*

Theodore Delevoryas

Willard D. Hartman

Keith S. Thomson

Alfred W. Crompton, *ex officio*

EDITOR: Jeanne E. Remington

ASST. EDITOR: Nancy A. Ahlstrom

Communications concerning purchase or exchange of publications should be addressed to the Publications Office, Peabody Museum of Natural History, Yale University, New Haven, Connecticut 06520, U.S.A.

Printed in the United States of America

CONTENTS

LIST OF ILLUSTRATIONS

ABSTRACT

Exposures of the Fox Hills Formation in Dewey, Ziebach and Corson counties, northwest-central South Dakota, include its historical type locality and constitute its type area. Here the formation consists of 300 to 350 feet of dominantly sandy, fossiliferous, marine and brackish-water strata gradational into the marine Pierre Shale below and the nonmarine Hell Creek Formation above. Since the pioneer work of Meek and Hayden, described herein, no thorough faunal or stratigraphic study has been made of the type Fox Hills. Together with contiguous outcrops in the Missouri Valley it constitutes the youngest marine Cretaceous (Maestrichtian) terrain in the western interior; what part of the Maestrichtian stage it represents is open to question. The type Fox Hills Formation is correlative with the lower part of the nonmarine Lance Formation in its type area in eastern Wyoming; this westward change to nonmarine strata of equivalent age takes place within fifty miles of the type Fox Hills.

The formation has distinctive lower and upper parts in the type area. In the lower part a basal clayey silt (Trail City Member) grades upward and laterally into a wedge-like sand body (Timber Lake Member) which pinches out westward within the type area. The upper part (Iron Lightning Member, new name) consists of two dominantly sandy, intergrading lithofacies and rests with sharp contact on the lower part. These two parts of the Fox Hills are products of different depositional regimes.

In the lower Fox Hills a number of concretion layers contain individually distinctive fossil assemblages, each of which apparently formed simultaneously throughout its extent. Together with other key beds these assemblage zones, which are believed to represent recurrent mass killings of natural settlements of benthic organisms, provide a framework for reconstruction of lithofacies and biofacies distribution at successive levels. This "time-lapse" method of environmental reconstruction reveals that a southwest-flowing, coastwise current was the major factor influencing sediment and faunal distribution in the lower Fox Hills. In the eastern two thirds of the type area the Trail City Member is largely biogenically-mixed clayey silt with several richly fossiliferous concretion layers forming assemblage zones in its lower part (Little Eagle lithofacies). Westward the member grades to dominantly thin-bedded silt and shale with mostly barren concretions (Irish Creek lithofacies). The lower part of the Timber Lake Member grades out westward into the Irish Creek lithofacies, but northeastward—beyond the type area—it gradually replaces most of the Little Eagle lithofacies. The populous, benthic settlements represented by the assemblage zones of the Little Eagle lithofacies formed only off the down-current end of the Timber Lake sand body, which advanced into the area from the northeast. Marked local changes in fossil associations reflect a variety of environments on and around the sand body itself; rich settlements like those of the lower Trail City formed on its deeper south end, their faunas becoming less diverse northward around its shallower axial part which supported a very restricted fauna in its high subtidal to intertidal environment.

The upper Fox Hills, or Iron Lightning Member, includes conspicuously thin-bedded sand, silt and shale with a sparse marine fauna (Bullhead lithofacies) and bodies of clayey, grayish-white sand with a brackish-water fauna (Colgate lithofacies). These two lithofacies, previously classed as members, cannot be consistently separated in the type area. The Iron Lightning Member rests on planed surfaces at several levels on the lower Fox Hills, rising steplike eastward. Upward and westward it passes into the lignitic clays and sands of the Hell Creek Formation. Iron Lightning sediments are those of a deltaic front advancing from the west, first filling the area shoreward of the Timber Lake sand body, then overstepping the body. Repetition of sands of the distributary system (Colgate) at more than one horizon in the delta-front sediments (Bullhead) suggests some fluctuation of sea level before the marsh, swamp and coastal plain deposits of the subaerial delta were deposited to form the lower part of the Hell Creek Formation. A number of different fossil associations of low diversity document the marine-nonmarine transition in the delta environments.

ZUSAMMENFASSUNG

Ausbisse der Fox Hills Formation in den Dewey, Ziebach und Corson Bezirken in nordwest zentral Süd-Dakota umfassen seine historische Typlokalität und bilden seinen Typgebiet. Die Formation besteht hier aus 300 bis 350 Fuss von vorwiegend sandigen, fossilienhaltigen Meeres- und Brackwasserschichten, welche allmählich in die untere marine Pierre Shale und in die obere kontinentale Hell Creek Formation übergehen. Seit der hierbeschriebenen Pionier-Arbeit von Meek und Hayden, war keine gründliche Studie der Fauna oder der Stratigraphie der Fox Hills im Typgebiet gemacht. Zusammen mit den angrenzenden Ausbissen im Missouri Tal bilden die Fox Hills das jüngste marine kretazische (Maastricht) Gebiet im zentralen Westen. Es fragt sich jedoch, welcher Teil des Maastrichts ist representiert. Die Fox Hills Formation im Typgebiet entspricht dem unteren Teil der kontinentalen Lance Formation in seiner Typgebiet im östlichen Wyoming; diese Veränderung nach Westen zur kontinentalen Schichten von entsprechendem Alter findet innerhalb fünfzig Meilen vom Typgebiet der Fox Hills statt.

Die Formation hat ausgeprägte untere und obere Teile im Typgebiet. Basaler, toniger Silt des unteren Teils (Trail City Member) geht nach oben und seitwärts in eine keilartige Sandbank (Timber Lake Member) über, welche nach Westen innerhalb des Gebiets auskeilt. Der obere Teil (Iron Lightning Member) besteht aus zwei vorwiegend sandigen, ineinander übergehenden Lithofazies und liegt sharf begrenzt am unteren Teil. Diese Zwei Teile der Fox Hills sind Ergebnisse verschiedener Ablagerungsvorgänge.

Mehrere Konkretionslagen in der unteren Fox Hills enthalten individuel underscheidende fossile Anhäufungen, von denen auscheinend jede im vollen Umfang gleichzeitig gebildet wurde. Diese Anhäufungszonen (Assemblage Zones), von denen man glaubt, dass sie ein wiederholtes Massensterben natürlicher Besiedelungen von benthischen Organismen representieren, bilden, zusammen mit den anderen kritischen Schichten, die Grundrahmen für die Rekonstruktion von Lithofazies und Biofazies in aufeinanderfolgenden Niveaus. Diese "Zeitspanne-Methode" der Umgebungsrekonstruktion zeigt, dass ein süd-westlichfliessender Küstenstrom der Hauptfaktor war, der die Ablagerung und die Faunaverteilung in den unteren Fox Hills beeinflüsste. Der Trail City Member in den östlichen Zwei-Dritteln des Gebiets besteht meistens aus biogenischem, tonigem Silt mit mehreren sehr fossilienreichen Konkretionslagen, welche die Anhäufungszonen des unteren Teils (Little Eagle Lithofazies) bilden. Nach Westen geht der Trail City in vorherrschend feinschichtlichen Silt und Schieferton meistens mit unfruchtbaren Konkretionen (Irish Creek Lithofazies) über. Der untere Teil des

Timber Lake Member geht nach Westen in die Irish Creek Lithofazies über, nach Nordosten aber, ausserhalb des Hauptgebiets, verdrängt er allmählich fast alle Little Eagle Lithofazies. Die dicht bevölkerten, benthischen Besiedelungen, welche die Anhäufungszonen der Little Eagle Lithofazies repräsentieren, wurden nur vor dem stromabwärts-gerichteten Ende der Timber Lake Sandbank gebildet, welche von Nordosten in dieses Gebiet vorrückte. Ausgeprägte lokale Veränderungen in fossilen Assoziationen zeigen, dass man auf der Sandbank und in Seiner Nähe viele verschiedene Umgebungen findet. Fruchtbare Besiedelungen, wie diese der unteren Trail City Ablagerungen, wurden im tiefer gelegenen Südende gebildet. Faunen wurden weniger manigfaltig näher zum flacheren Achsialteil der Sandbank; eine sehr beschränkte Fauna lebte in seiner Flutzone- und unmittelbar darunter.

Die obere Fox Hills, oder Iron Lightning Member, umfasst auffallend feinschichtlichen Sand, Silt und Schieferton mit spärlicher mariner Fauna (Bullhead Lithofazies) sowie Schichten von tonigem, gräulich-weissem Sand mit Brackwasser Fauna (Colgate Lithofazies). Man kann nicht immer diese zwei Lithofazies im Typgebiet ausscheiden. Der Iron Lightning Member liegt an planierten Oberflächen an mehreren Niveaus der unteren Fox Hills und wächst stufenartig ostwärts. Nach oben und westwärts geht er in den lignitischen Ton und Sand der Hell Creek Formation über. Die Iron Lightning Ablagerungen sind Produkten einer Deltafront von Westen, die zuerst das küstliche Gebiet der Timber Lake Sandbank ausfüllten und nachher dieselbe überwuchsen. Wiederholte Vorkommen von Sand der Deltaarmen (Colgate) auf mehreren Horizonten in den Deltafrontablagerungen (Bullhead) suggerieren ein gewisses Schwanken des Meeresspiegels ehe Sumpf und Flussablagerungen der Deltaebene abgelagert wurden den unteren Teil der Hell Creek Formation zu bilden. Mehrere fossile Assoziationen von geringer Mannigfaltigkeit beurkunden den marine-kontinentale Übergang in den Deltaumgebungen.

ТИПОВАЯ ФОРМАЦИЯ ФОКС ХИЛЛС, МЕЛ (МААСТРИХТ), ЮЖНАЯ ДАКОТА

КАРЛ М. ВААГЕ

РЕЗЮМЕ

Обнажения формации Фокс Хиллс (Fox Hills Formation) в графствах Дюи (Dewey), Зибах (Ziebach) и Корсон (Corson), в северозападно-центральной Южной Дакоте, включают ее историческую типовую местность и сочиняют ее типовую область. Здесь эта формация состоит из 300-350 футов преобладающе песчанистых, содержащих окаменелости, морских и отложенных из солоноватой воды слоев, переходящих в морскую формацию Пиер Шейл (Pierre Shale Formation) внизу и континентальную формацию Хелл Крик (Hell Creek Formation) наверху. От времени пионерской работы Мика (Meek) и Хейдена (Hayden), описанной здесь, никакая фаунальная или стратиграфическая студия типовой Фокс Хиллс не была сделана. Вместе с соседними обнажениями в долине Миссури она сочиняет младшую морскую меловую (маастрихтскую) серию пластов внутренней части Запада; какую часть маастрихтского яруса она представляет — открытый вопрос. Типовая формация Фокс Хиллс коррелирует с нижней частью континентальной формации Ленс (Lance Formation) в ее типовой области в восточном Вайоминге. Эта перемена в направлении запада в континентальные пласты эквивалентной старости совершается не дальше чем 50 миль от типовой местности формации Фокс Хиллс.

В типовой области у формации отличаемы нижняя и верхняя часть. В нижней части базальный глинистый тонкозернистый песок (член Трейл Сити — Trail City Member) переходит постепенно вверх и в сторону в клинообразное песчанное тело (член Тимбер Лейк — Timber Lake Member), исклиняющее к западу в пределах типовой области. Верхняя часть (член Айрон Лайтнинг — Iron Lightning Member, новое название) состоит из двух преобладающе песчанистых, интерградированных литофаций и залегает с острым контактом на нижней части. Эти две части формации Фокс Хиллс — результаты различных режимов отложения.

В нижней формации Фокс Хиллс многие слои конкреций содержат индивидуально отличительные накопления окаменелостей, каждое из которых кажется образованным одновременно во всем его протяжении. Вместе с другими опорными пластами эти зоны накопления, для которых считается, что они отвечают повторяющимся массовым вымираниям природных поселений бентонных организмов — дают нам рамы для реконструкции распределения литофаций и биофаций в последовательных горизонтах. Этот метод реконструкции среды в коротких промежутках времени обнаруживает, что береговой ток в направлении юго-запада был важным фактором, влияющим на распределение отложений и фауны в нижней части формации Фокс Хиллс. В восточных двух третях типовой области член Трейл Сити является в значительной части биогенно-перемешанными глинистыми тонкозернистыми песками; несколько конкреционных слоев, богатых окаменелостьих, образуют зоны накопления в его нижней части (литофация Литл Игл — Little Eagle). К западу член постепенно переходит в преобладающе тонкослоистые тонкозернистые пески и сланцы, конкреции в которых по большей части без окаменелостей (литофация Айриш Крик — Irish Creek). Нижняя часть члена Тимбер Лейк постепенно переходит к западу в литофацию Айриш Крик, но к северо-востоку — вне типовой области — он постепенно сменяет бо́льшую часть литофации Литл Игл. Населенные бентонные поселения представленные зонами накопления литофации Литл Игл образовались только около того (глубокого) конца песчанного тела Тимбер Лейк, в направлении которого течение совершалось. Наступление песчанного тела в рассматриваемую область было с северовостока. Заметные местные изменения в ассоциациях окаменелостей отражают разнообразие сред на и в окрестности песчанного тела. Богатые поселения подобные тем нижнего Трейл Сити образовались на его глубжем, южном концу; их фауны становились менее разнообразным к северу, вблизи его менее глубокой, осевой части. В приливной и подприливной среде последней жила очень ограниченная фауна.

Верхняя часть Фокс Хиллс, или член Айрон Лайтнинг, включает заметно тонкослоистый песок, тонкозернистый песок и сланец, с редкой морской фауной (литофация Буллхэд — Bullhead) и телами глинистого, серо-белого песка с фауной солоноватой воды (литофация Колгейт — Colgate). Эти две литофации, прежде классифицированые членами, не могут быть непротиворечиво выделены в типовой области. Член Айрон Лайтнинг залегает на сровненных поверхностях на нескольких уровнях в нижней Фокс Хиллс, поднимаясь степенообразно к востоку. Вверх и к западу он сменяется лигнитовыми глинами и песками формации Хелл Крик. Отложения члена Айрон Лайтнинг — отложения дельтового фронта, наступление которого было с запада и который сначала выполнил простор между песчанным телом Тимбер Лейк и берегом, а потом распространился выше Тимбер Лейка. Повторение песков распределительной системы (Колгейт) на нескольких горизонтах в отложениях дельтового фронта (Буллхэд) говорит в пользу флюктуаций морского уровня прежде чем отложения болот и береговой равнины образовали нижнюю часть формации Хелл Крик. Многочисленные различные ассоциации окаменелостей низкого разнообразия документируют сменение морских условий континентальными в дельтовых средах.

1. INTRODUCTION

SCOPE

The dominantly marginal marine and brackish water deposits that compose the Upper Cretaceous Fox Hills Formation were first studied and named in the Missouri Valley region of the northeastern Great Plains, chiefly in the central Dakotas. With few exceptions, little has been added to our knowledge of the formation in this area. Both here and elsewhere in the northern Great Plains and Rocky Mountain Region the formation has rarely been a specific subject of study, our rather superficial knowledge of it coming chiefly from regional and economic geological investigations of areas in which it happens to crop out. Study of the type area of the Fox Hills in South Dakota was begun with the objective of providing a useful standard of comparison for the latest Cretaceous marine strata of the western interior, for it was apparent from existing work that the Missouri Valley Fox Hills is the youngest marine Cretaceous in that region.

The field study was not far along before the unique opportunity to use the area for detailed environmental and ecological studies became evident. Few other terrains of interior Late Cretaceous rocks contain as complete a fossil record of the marine-continental transition as the type area of the Fox Hills Formation. From the shallow-water marine beds of its basal member, through its upper brackish-water beds and into the overlying fresh-water deposits of the lower Hell Creek Formation, fossils are locally common in most of the facies and in some they are exceptionally abundant. In the continuous outcrop of the Fox Hills that extends west from the type area around the Black Hills and south into eastern Wyoming (Fig. 1) the formation is scantly fossiliferous except in the Lance Creek area of Wyoming; here fossils are common in a number of facies but are neither as abundant nor as diverse in faunal associations as in the type area. Elsewhere in the northern Great Plains the record is similar, featuring extensive, scantly fossiliferous outcrop of the Fox Hills, or its litho-genetic equivalent, and a very few, localized areas with relatively abundant fossils. In general, brackish-water faunas are more consistently present in the Fox Hills out-crops than shallow-water marine faunas; rich accumulations of the latter are, indeed, local rarities. As will be seen, even within the exceptionally fossiliferous type area the abundant benthic marine faunas are restricted in distribution to local areas.

Preliminary stratigraphic studies revealed marked patterns in stratigraphic and geographic distribution among the rich faunas, relative both to lithofacies and to associations of different invertebrate genera, indicating that fossil distribution in the type Fox Hills reflected the original distribution of organisms. Consequently, the project was oriented to take advantage of the opportunity that these faunas offered for detailed ecological and environmental studies of the marginal deposits of an epiconti-nental sea. This report, the necessary initial step toward the broader aim, deals with the details of the local stratigraphy and the more conspicuous features of the fossil distribution. Inclusion of a fairly detailed historical summary seemed proper since this is a re-examination of one of the original five units of Meek and Hayden's initial classification of the interior Cretaceous. In addition this study will serve as a foundation for another phase of the project—the systematic studies of the invertebrates necessary to modernize the antiquated taxonomy, a prerequisite for any valid ecological work. A study of the Fox Hills bivalves, the most helpful group in

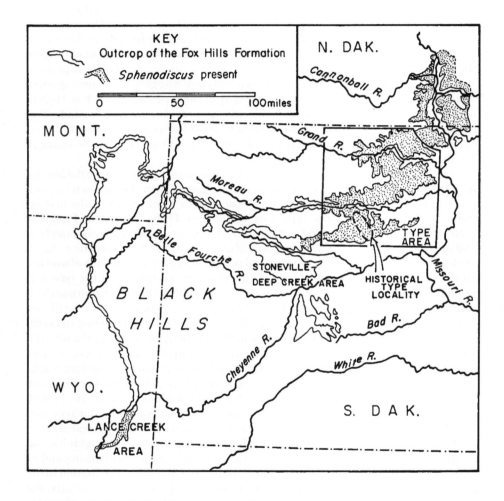

FIG. 1. Outcrop of the Fox Hills Formation in South Dakota and adjacent areas.

deciphering environments, has been completed by Dr. Ian Speden and is being prepared for publication. Speden's work will alter some of the fossil names but this report, because of its earlier appearance, must follow existing nomenclature, which is mostly that of Meek and Hayden or published changes of it.

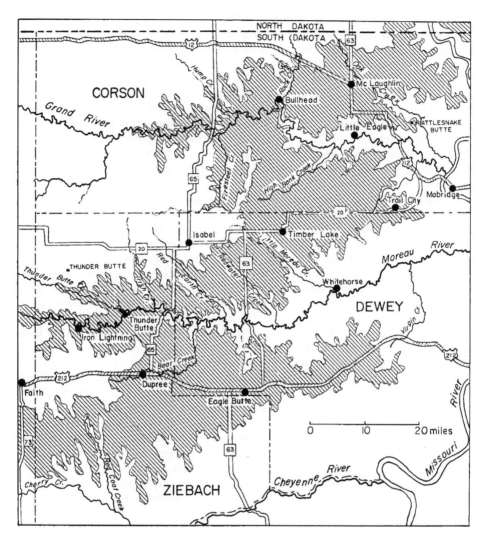

FIG. 2. General index map of Corson, Dewey and Ziebach Counties, South Dakota. The type area of the Fox Hills Formation includes all of its outcrop (cross hatched) in the three counties except that northeast of Oak Creek. (Missouri valley area shown prior to flooding by Oahe Reservoir).

FIELD WORK

To overcome deficiencies of exposure and stratigraphic coverage the historic Meek and Hayden type locality of the Fox Hills had to be broadly interpreted and the type area (Fig. 1) enlarged in geographic scope. The advantages of this type area as a field laboratory for studies of various aspects of a shallowing epicontinental sea are impaired somewhat by rather poor bedrock exposures. Unlike equivalent strata in the western plains and Rocky Mountains the Upper Cretaceous strata of the Missouri Valley are poorly consolidated, resembling in this respect the Cretaceous deposits of the Atlantic Coastal Plain. The type area of the Fox Hills includes rolling prairie uplands dissected by drainage systems of eastward-flowing rivers tributary to the Missouri (Fig. 2). Both the intricately dissected breaks along the major rivers and the scattered buttes on the upland are largely grass covered. The slight regional dip northwestward into the Williston basin results in a step-like pattern of outcrop so that successively higher levels of the Fox Hills and succeeding formations underlie the upland divides to the west. This feature tends to separate exposures of the lower part of the formation from those of the upper part except at a very few places in the type area. Nowhere in the area is a complete section of the entire Fox Hills Formation found in a single exposure.

Field work in the type area was carried on intermittently for a total of about seven months from 1957 through 1964. Standard field methods were used in as much detail as was practical. The localities were related to one another by lateral tracing and matching of sequences of key beds. From individual measured sections in closely adjacent areas of outcrop, composite sections were constructed for the exposed sequence of Fox Hills and adjacent strata. No attempt was made to prepare a detailed geologic map of the area as it is currently being done at a scale of one mile to the inch by the South Dakota Geological Survey; but some mapping of key beds and contacts was done where control was needed for relating measured sections and this has led to revision of the pattern of Fox Hills outcrop found on the State geologic map (Fig. 2 and 15). With few exceptions the position of each lithology sample or fossil collection was fixed relative to a measured section and the fossils of individual concretions and local lenses treated as separate collection lots. The bulk of the detail on which the study is based comes from more than 300 principal localities; only those localities specifically noted in the text are shown on the locality map (Fig. 15).

No comprehensive sedimentary studies were made of any of the Fox Hills units. The qualitative descriptions given are based on field comparisons, under a hand lens, with a standard set of grade size samples. These and other field observations were supplemented by size analyses of about 20 spot samples and examination of 30 thin sections as well as laboratory observations on numerous rock specimens. The terminology of grade sizes follows the Wentworth scale throughout (see Figs. 8 and 9).

ACKNOWLEDGEMENTS

The present study has been supported chiefly by National Science Foundation grants G-5657, G-18674; a third grant, GP-4467, is presently supporting continuation of the project in other areas. I am grateful to the Foundation for its advice and patience, as well as for the financial aid that made the work possible. This Bulletin is published with the aid of N.S.F. Publication Grant No. GN-528. Field work in 1957 and field transportation costs throughout much of the project have been supported by the Charles Schuchert Fund of Yale University.

From the spring of 1962 until he completed his doctoral study at Yale in November 1964, I had the assistance and close collaboration of Dr. Ian G. Speden of the New

Zealand Geological Survey—an association both helpful and enjoyable. His contributions to this work are many.

The project has benefited from the assistance of a number of students. Particularly rewarding was the field assistance and companionship of James F. Mello and Leonard Radinsky in 1959, Speden in 1962, and Jeremy Reiskind in 1964, then graduate students at Yale University. Peter Bretsky, Martin Buzas and Speden contributed significantly in laboratory studies of Fox Hills sediments and fossils. Edward Gilfillan, Roger Podewell and Peter Scholle then undergraduates at Yale, contributed to both field and laboratory work. Other field assistants during their undergraduate years included Michael Price (Yale), Leo Hickey (Villanova), and Jonathan Waage (Princeton); Sandy Harris and Peter Hilgeford, then secondary school students, also served as field assistants.

Dr. William A. Cobban, U.S. Geological Survey, generously loaned me sections that he measured years ago in the type area of the Fox Hills. Many of these exposures are now obscured and the otherwise inaccessible information has proved a valuable addition to the work. Helpful information on the local stratigraphy of the Pierre Shale was furnished by James F. Mello from his studies of this formation and its microfaunas (unpublished, 1962).

From the beginning I had the cooperation of Dr. Allen F. Agnew, State Geologist of South Dakota until 1963, and his successor Dr. Duncan J. McGregor, and have benefited from discussion and field conferences with Agnew and other members of the South Dakota Geological Survey, in particular Earl J. Cox, Allen Lange, Wayne Pettyjohn and Robert E. Stevenson. I also wish to express my appreciation to many of the people in and around Timber Lake, South Dakota, whose friendship and generous helpfulness made this base of operations seem like a second home.

For their helpfulness, and generosity in furnishing me with data on Meek and Hayden I am indebted to Dr. F. M. Fryxell and Dr. J. V. Howell, who provided me with copies of pertinent Hayden letters, a sketch map of his, and field notes of Meek, all of which helped greatly in clarifying their knowledge and interpretation of the Fox Hills Formation. I am also indebted to Dr. William H. Goetzmann of the University of Texas for rewarding discussions of early western explorations; and to Dr. Archibald Hanna, Jr., for his help in finding useful material in the Yale Western Americana Collection, of which he is Curator.

I am particularly indebted to Jeanne E. Remington, and to Drs. Cobban and Speden, for their helpful editing and reviews of the manuscript. Preparation of the illustrations and typescripts was accomplished with the help of the Peabody Museum and Geology Department staff members Judy Chaney, Martha Dimock, Roberta French, Louise Holtzinger, John Howard, and Solene Roming, to all of whom I am most grateful.

2. THE TYPE FOX HILLS

FOX HILLS FORMATION DEFINED

The Fox Hills Formation made its debut in geological literature as Formation No. 5, the sandy marine deposits at the top of the sequence, in the first formal classification of western interior Cretaceous strata by Meek and Hayden (1856a, p. 63). Subsequently they replaced numbers with names, noting that the Fox Hills Group was gradational below with the marine shale of the Fort Pierre Group and gradational above with the brackish- and fresh-water beds of the Great Lignite Group, which they considered Tertiary (Meek and Hayden, 1861, p. 419, 427). Their acquaintance with and concept of the formation are reviewed below under "History of study". Subsequent to this early work Hayden and his contemporaries applied the name Fox Hills to the sandy transition beds at the top of the marine Cretaceous throughout much of the northern Great Plains and Rocky Mountain region.

In the years following this pioneer work, names other than Fox Hills were given to these sandy transition beds in some places because many geologists were reluctant to use this name for deposits that were obviously older than those of the type area in central South Dakota, even though they recognized that these deposits were genetically equivalent to the type Fox Hills. Subsequent attempts to reverse this trend toward the proliferation of names were rare (Bartram, 1937) and not successful. Consequently, the name Fox Hills became restricted in common usage (Cobban and Reeside, 1952) largely to areas of outcrop in North and South Dakota, eastern Montana, eastern and parts of central Wyoming, and eastern Colorado south to the Colorado Springs area. This distribution has no paleogeographic significance since within the area encompassed by the name the formation varies both in lithology and in age. Its approximate restriction to the northern Great Plains east of longitude 106° is largely an artifact of: 1) the outcrop pattern, 2) the history of study, and 3) the individual preference of investigators in applying stratigraphic nomenclature.

Like most of the pioneer formations the Fox Hills has had a fairly complex history of stratigraphic definition. This can be summarized here by examining briefly the four different criteria used in attempts to establish uniform boundaries for the formation, namely: lithology, genesis, unconformity, and age. The criterion of unconformity was eliminated by the work of Dobbin and Reeside (1929) on the Fox Hills—Lance boundary, and the gradational nature of both its upper and lower contacts, originally recognized by Meek and Hayden, is now generally accepted. The time-transgressive nature of the Fox Hills is also an accepted fact: consequently, the criterion of age—except where used in a supplementary sense too broad to be pertinent for local delimitation—cannot be employed in its definition.

A review of existing definitions of the Fox Hills that are based on lithology and/or genesis reveals that no precise delimitation of the formation holds for the entire outcrop area or even for a very large part of it. Uniformity of sequence over any appreciable area simply is not a characteristic of marginal marine deposits. Pronounced local differences in lithology and genesis are common in the Fox Hills and require separate local definitions based on easily recognizable lithologic features that afford mappable contacts. For regional description the Fox Hills can be defined only in general terms as a dominantly sandy sequence of marginal marine and brackish-water sediments variable in thickness and lithology and gradational below with the gray marine clay shale of the Pierre Shale and its equivalents and above with the dominantly continental sands and clays of the Lance, Hell Creek and equivalent formations.

In regard to the formal name of the Fox Hills, the usage Fox Hills Formation is preferable to Fox Hills Sandstone and should replace it. I do not think this minor semantic revision merits any argument here. The detailed historical summary that follows reveals that Fox Hills Formation was the original usage in the type and adjacent areas and demonstrates that the later introduction of the term Fox Hills Sandstone brought with it a change in concept that is not in harmony with its variable lithofacies. For consistent usage throughout its area of outcrop, Fox Hills Formation is the most useful name for this lithologically heterogeneous unit.

TYPE LOCALITY AND TYPE AREA

Meek and Hayden noted (1861, p. 427) that the Fox Hills beds were "... most distinctly marked at Fox Hills between Cheyenne and Moreau rivers . . ." and ever since, an ill-defined part of the Cheyenne-Moreau divide has been cited as the type locality.

Today U.S. Highway 212 traverses the Cheyenne-Moreau divide (see Fig. 2) and travelers driving it westward from the Missouri see very little resembling a ridge, or hills, in the area of Fox Hills outcrop. Yet the literature on the formation gives the impression that the type exposures are to be found in a well-defined ridge or range of hills. As early as 1910 James E. Todd (1910, p. 18), the first State Geologist of South Dakota, called attention to this apparent geographic misnomer in his work on the geology of the Moreau and Grand River drainages. He reasoned that the voyageurs and early exploring parties looking westward from the Missouri, or northward from the valley of the Cheyenne, mistook for a ridge the distant, abrupt edge of the tableland forming the Cheyenne-Moreau divide and there is ample support for this contention in the records of exploration from Lewis and Clark on. Even the first geological reconnaissance of note in this area—John Evans' trip to the White River Badlands for David Dale Owen—promoted this idea. Evans' understandably inaccurate sketch map (Owen, 1852, following plate 15) of the setting of the badlands shows two prominent ranges of hills: the Black Hills on the west, and the smaller Fox Hills on the east lying between the Cheyenne and Moreau Rivers, east of Cherry Creek. To my knowledge, this is the earliest published map bearing the name Fox Hills.

Lieut. G. K. Warren, during his first expedition in the West in 1855, solicited a number of sketch maps of Upper Missouri country from mountain men (Goetzmann,

1959, p. 410); these are now preserved in the G. K. Warren Papers at the New York State Library in Albany. One sketch by Michael Desomet bears the following notation between the Cheyenne and Moreau Rivers, "Prairie Fox Ridge, 20 miles from Shayen's mouth." The south-facing bluff of the Cheyenne-Moreau divide is just about that distance northwest of the mouth of the Cheyenne. Warren's sketches also include two from Hayden, one of which is a crude sketch of a traverse across the Cheyenne-Moreau divide. This shows Fox Ridge as a narrow bluff running roughly east-west approximately 15 miles north of the Cheyenne River, corresponding very well with the "breaks" on the Fox Hills Formation that form the south edge of the Cheyenne-Moreau divide. These breaks consist of a zone of rounded, somewhat intricately dissected but grass-covered topography 1½ or 2 miles wide, within which there is a drop in elevation to the south of about 200 feet. The actual divide lies just to the north of these breaks. Knowing from his sketch approximately where Hayden crossed the divide it is possible to say that the type locality lies within the belt of Fox Hills outcrop along the south edge of the Cheyenne-Moreau divide between S. Dak. Highway 63 and the road south from Dupree that parallels it to the west. But in Hayden's day the rigorous definition of type localities had not yet become a part of stratigraphic ritual and it seems best to qualify this type locality after-the-fact by calling it the historical type locality (Fig. 1).

Subsequently Fox Hills, or Fox Ridge, came to include far more territory than the Cheyenne-Moreau divide of Evans or the breaks so labelled by Hayden. In describing his Formation No. 5 at the point where its outcrop crosses the Missouri north of what is now the South Dakota—North Dakota state line, Hayden (1857a, p. 113-114) wrote:

> Here it forms an extension of what is called Fox Ridge, a series of high hills having a northeast and southwest course, crossing the Missouri River into Minnesota at this point. Its northeastern limits I have not ascertained. In its southwestern extension it continues for a considerable distance nearly parallel with the Missouri, crosses the Moreau River about thirty miles above its mouth, then forms a high dividing ridge between the Moreau and Shyenne Rivers, at which locality it first took its name.

This description is part of the text accompanying Hayden's first geologic map of the Upper Missouri country. On the map the words "Fox Ridge" extend northeastward across the Cheyenne-Moreau and Moreau-Grand divides.

Fox Ridge, in its broadest usage, was the name applied to the bluffs forming the southern and eastern edges of the dissected tableland underlain by the relatively flat-lying, sandy strata of Formation No. 5. This broad use of the name Fox Ridge persisted until the early 1860's, about which time the term appears to have become restricted again to the eastern part of the Cheyenne-Moreau divide.

With the coming of white settlement in the area during the present century, the name Fox Ridge became attached to a specific topographic feature—a group of buttes lying just south of U.S. Highway 212 between 20 and 25 miles west of Faith in northern Meade County. This Fox Ridge is accepted by the Board of Geographic Names and appears on all recent maps of the region, scale permitting. Unfortunately, this Fox Ridge was never even so much as a part of the Fox Ridge of Hayden, his predecessors or his contemporaries. The buttes of the modern Fox Ridge are Tertiary strata that rest on a broad upland underlain by the Hell Creek Formation, and are 30 to 40 miles west of the terrain to which the name was originally applied. For a time, during the 1920's and 1930's, there was even a small community post office on Highway 212 near the buttes that was also called Fox Ridge; although abandoned prior to World War II, it still appears as a settlement in some atlases and road maps. The modern Fox Ridge has misled its share of geologists, and it should be clearly understood that this is *not* the Fox Ridge of Hayden and that it lies to the west of both the historical type locality of the Fox Hills and the broader type area as defined in this report.

None of the few outcrops in the area of the historical type locality afford an adequate type section of the Fox Hills Formation in terms of modern stratigraphic work. Only scattered parts of the lower marine beds of the formation—particularly those of the sandy Timber Lake Member—crop out. The beds of the upper half of the formation come in along the divide to the west and northwest. Exposures nearest to the historical type locality in which reasonably adjacent outcrops afford a composite section of the formation lie about 10 miles west of a road south from Dupree in the vicinity of Red Coat Creek. But here lateral change in facies westward has eliminated the fossiliferous marine sandstone unit (Timber Lake Member) that is the most characteristic part of the formation in the historical type locality, where it holds the bluff that Hayden called Fox Ridge. This sandstone unit not only underlies the east end of the Cheyenne-Moreau divide, but also extends northward, forming the east end of the Moreau-Grand divide and the tableland extending northeastward from the Grand River to the mouth of the Cannonball River and across the Missouri into Emmons County, North Dakota. In other words, it is largely this characteristic sand unit that supports the rolling, steep-sided upland, the edges of which Hayden included in his extended application of the name Fox Ridge.

Although Hayden's type locality along the south side of the Cheyenne-Moreau divide suffices as an historical marker a broader area of outcrop is required for adequate stratigraphic reference to the typical Fox Hills succession. The reference area proposed, and subsequently referred to here as the type area, includes the Fox Hills outcrop in Dewey and Ziebach Counties and in Corson County south of the valley of Oak Creek, South Dakota (see Fig. 2). This area includes Hayden's type locality of the formation and also the type localities of all three members of the Fox Hills Formation recognized in this study. So large a type area is necessary to include: 1) adequate exposures of all the members and the upper and lower contacts of the formation with the Hell Creek and Pierre Formations respectively; and 2) sufficient lateral extent of outcrop to demonstrate the nature of the pronounced lateral changes in lithofacies to the west and the more gradual change in both lithofacies and biofacies to the north and northeast. As will be seen, Hayden covered a large part of this area and the two most critical studies of the Fox Hills since his work (Todd, 1910; Morgan and Petsch, 1945) were both made within it.

REGIONAL RELATIONSHIPS

The Fox Hills Formation varies considerably in both age and character within the continuous belt of outcrop (Fig. 1) that extends from the Missouri Valley area westward around the north end of the Black Hills then southward to the Lance Creek area of Niobrara County, Wyoming. The formation becomes older as it is traced into the area north of the Black Hills from either end of the outcrop belt. This change in age is inferred from two criteria. One is the position of the base of the Fox Hills Formation relative to the successive zones of baculites in the underlying Pierre Shale; the highest of these is the Range Zone of *Baculites clinolobatus* Elias—the last of the indigenous interior baculites (Cobban, 1958, p. 114). The other is the distribution of the ammonoid genus *Sphenodiscus* in the Fox Hills beds; this is shown in Fig. 1.

The restricted distribution of *Sphenodiscus* in the outcrop belt is due in part to change of facies as well as to change in age. *Sphenodiscus*, whose stratigraphic range in the interior region begins in the *Baculites grandis* Range Zone (Cobban and Reeside, 1952, pp. 1020-1021), would probably be present in parts of the outcrop belt other than the two areas shown in Fig. 1 had the proper near-shore marine conditions prevailed in these parts during the deposition of the Fox Hills sediments. Consequently, the more reliable criterion in judging the relative age of the Fox Hills is the stratigraphic relationship to the occurrence of *B. clinolobatus*.

A brief stratigraphic summary of the Fox Hills in western South Dakota and adjacent parts of neighboring states will serve to put the type Fox Hills within the context of regional stratigraphy.

MISSOURI VALLEY AREA

The Fox Hills exposures in the type area together with the contiguous outcrop in adjacent parts of the Missouri Valley of North and South Dakota include the youngest ammonite-bearing marine rocks in the western interior, and record the last major stand of the interior Cretaceous sea. With some exceptions the stratigraphy of the formation in this entire area is similar to that of the type area which forms, approximately, its southwestern half. The eastern limit of the Fox Hills is an irregular boundary just east of the Missouri River where the formation disappears beneath Pleistocene cover.

The Fox Hills Formation consists of about 300 to 350 feet of dominantly silty and sandy beds grading downward into the Pierre Shale and upward into the continental beds of the Hell Creek Formation. The base of the formation is separated from the upper beds of the *Baculites clinolobatus* range zone by about 250 feet of relatively unfossiliferous shale that includes the uppermost part of the Mobridge Member and overlying Elk Butte Member of the Pierre Shale.

The lowest part of the Fox Hills consists of gray, clayey silt (Trail City Member) superficially similar to the underlying Pierre Shale on fresh cuts, but distinctly more silty. Concretion layers in the clayey silt locally contain a very rich marine molluscan fauna. Over much of the area the clayey silt grades upward into a dominantly fine-grained, greenish-gray dirty sand that weathers yellowish-orange (Timber Lake Member). This unit is a complex sand body the base of which drops in the section northeastward at the expense of the underlying clayey silt; the body terminates abruptly to the southwest within the type area. It contains an abundant marine fauna in its lower part and a much less diverse one in its upper part. Together the clayey silt and sand units form a distinctive, shallow-water marine, lower part of the Fox Hills Formation.

The upper part of the formation is made up of two intergrading lithofacies currently classed as members: a conspicuously thin-bedded shale, silt and sand lithofacies (Bullhead Member) commonly referred to as "banded beds", and bodies of fine- to medium-grained, dirty, clayey sand that weather grayish-white (Colgate Member). Locally the Bullhead banded beds have a sparse, specialized marine fauna and the Colgate sands carry an abundant if not diverse brackish-water fauna. The two closely related lithofacies of the Upper Fox Hills grade upward, and apparently westward into the lignitic clays and sands of the continental Hell Creek Formation. The Colgate sands are most conspicuous at the top of the formation but other lenses of the same lithofacies occur at lower levels within the banded beds of the Bullhead and also in the lower part of the overlying Hell Creek.

The ammonoid *Sphenodiscus lenticularis* (Owen) and a large scaphite *Discoscaphites nebrascensis* (Owen) that generally occurs with it are present in the Missouri Valley area, but are also found in other parts of the Fox Hills outcrop of the northeastern Great Plains. However, the Missouri Valley Fox Hills contains a variety of other scaphites not known elsewhere in the western interior. Principally on the basis of these species, which are of shorter range than the ubiquitous *Sphenodiscus*, and the considerable distance of the formation above the *Baculites clinolobatus* Range Zone, the Fox Hills fauna of the Missouri Valley is considered the youngest marine Cretaceous fauna in the interior region, excluding, of course, minor marine tongues with very small faunas found in the overlying Hell Creek Formation of the Missouri Valley area.

WESTERN SOUTH DAKOTA

The little that is known about the descent in stratigraphic section of the Pierre-Fox Hills contact in western South Dakota indicates that the initial change west of the Missouri Valley area is abrupt both in age and facies. The Fox Hills crops out over a large area in northern Meade County less than 50 miles west of the type area and contains a sequence of beds completely different from that of the type Fox Hills. Searight (1934), in a study around Stoneville in Meade County, separated as Fox Hills a dominantly sandy sequence of beds about 450 feet thick that include a few brackish-water fossils, some plant remains and a thin zone of coal-bearing beds. This sequence grades abruptly downward into Pierre Shale which contains near the top "... *Tardinarca (Pseudoptera) fibrosa* (Meek and Hayden), and a large baculite, probably *Baculites grandis* Hall and Meek" (Searight, 1934, p. 4). Pettyjohn (1967, p. 1361) has shown the baculite in question is *B. clinolobatus* and that it occurs 20 feet below the base of the Fox Hills in that area. The significant fact is that the entire marine sequence of the Missouri Valley Fox Hills and probably most of the upper 250 feet of the underlying Pierre Shale grade westward, within 50 miles, into a sandy sequence of predominantly brackish- and fresh-water beds in which none of the members of the type area are discernible. The nature of this lateral change is largely hidden in the relatively poor exposures between the two areas, but its beginnings are apparent in the southwestern part of the type area of the Fox Hills.

South across the Cheyenne River from the Fox Hills outcrop in Meade County is another area of equally peculiar Fox Hills situated north of the White River badlands around Deep Creek in northeastern Pennington and western Haakon Counties. Very little is known of the Fox Hills in this area. The uplands around Deep Creek are underlain by fairly well-indurated sand layers carrying a sparse shallow-water marine fauna; these are locally succeeded by a lithofacies very similar to the banded beds in the upper part of the type Fox Hills.

Together, the Fox Hills exposures in northern Meade County and the Deep Creek area include a highly complicated assortment of brackish- and probably fresh-water beds that contain lithofacies, such as the Stoneville lignite beds, that are unique to the formation in the northeastern Great Plains. Until work in progress reveals more about the nature of the Fox Hills in these areas, it will be convenient to consider them together as a distinctive phase of the Fox Hills that is referred to here as the Stoneville-Deep Creek complex (Fig. 1).

From the Stoneville-Deep Creek complex northwestward around the Black Hills rim and into central Carter County, Montana, the Fox Hills-Pierre contact descends in the section across several of the baculite zones in the upper Pierre, but the details are not known. The characteristic feature of Fox Hills stratigraphy in this area is the relatively sharp change from the marine shale of the Pierre to sandy, brackish- and fresh-water deposits with very few, if any, sandy beds containing marine fossils. Also characteristic in this sandy, largely nonmarine series of beds is the difficulty of finding a satisfactory contact between the Fox Hills and the Hell Creek. Consequently, there is a wide divergence of opinion on the thickness of the Fox Hills in the area.

EASTERN WYOMING

Considerably more is known about the details of Fox Hills chronology along the Wyoming side of the Black Hills than is known for western South Dakota. In Niobrara County the Fox Hills contains *Sphenodiscus* in and for a short distance north of the Lance Creek area, and the underlying Pierre Shale contains *Baculites clinolobatus* (Cobban, 1958, p. 114). Robinson, Mapel and Cobban (1959) have demonstrated that the Pierre-Fox Hills contact intersects progressively older baculite zones as it is traced northward through eastern Wyoming into the southeast corner of Montana.

Within the Fox Hills, however, the only truly abundant marine faunas found in eastern Wyoming occur at the end of the outcrop belt in the Lance Creek area. Northward, as on the eastern side of the Black Hills, marine species appear to be progressively less common in the Fox Hills. It is interesting that the gross stratigraphic features of the Fox Hills in the Lance Creek area are roughly similar to those of the Fox Hills in its type area. The upper 75 feet or more of the Pierre Shale in the Lance Creek area is sandy and resembles the lower clayey silt beds (Trail City Member) in general lithology, although its concretions are not as abundantly fossiliferous. Above this, the lower part of what is locally called the Fox Hills is chiefly massive sandstone like that in the Timber Lake Member and contains similar concretions with *Sphenodiscus* and other fossils. The massive sandstone is succeeded by a distinctive unit of thin-bedded sandstone with minor sandy shale which is overlain by massive white sands (locally called the Colgate Member) in the upper part of the formation. At one place a lens of banded beds identical to the Bullhead Member of the type area is associated with these massive white sands.

Summary

The dominantly marine areas of Fox Hills with comparable lithologic sequences that characterize the Missouri Valley and Lance Creek ends of the outcrop belt suggest a more gradual change from marine to continental conditions than is found in the intervening parts of the outcrop and may mark relative stability of the strand line. Only in these two areas does an appreciable part of the Fox Hills Formation appear to have been deposited under what can be called, for the interior Cretaceous sea, normal marine conditions. The Fox Hills Formation of the Missouri Valley is somewhat younger than that at Lance Creek. The intervening Stoneville-Deep Creek complex includes beds equivalent to both and played a critical role in the geography of the late phases of the interior Cretaceous sea in the northeastern Great Plains. Although little is known about the details of its structure, it most probably represents a persistent area of deltaic sedimentation.

3. HISTORY OF STUDY

INTRODUCTION

The stratigraphic results of Meek and Hayden's early work on the geology of the Upper Missouri country have come down to us chiefly in the form of a grand generalization—their classification of the Cretaceous and Tertiary sequence. The Fox Hills Formation, which has pioneer status in western stratigraphy as Formation No. 5 of the original classification, was studied during the earliest of Hayden's explorations in what was then Nebraska Territory. Few of these explorations were primarily geologic in nature, for Hayden had no funds of his own and moved around the country by joining any group of travelers or explorers he could persuade to support his scientific work. Consequently, what outcrops he visited, how thoroughly he studied them and where he collected fossils were determined by a variety of interacting economic, military and geographic factors, as well as by whim and weather. Separation of the geologic interpretations of Hayden and Meek from the circumstances of exploration would lessen both the understanding and appreciation of their contributions. For this reason much of what follows is a historical account of Meek and Hayden's investigation of the upper part of the Cretaceous section.

The published works of Meek and Hayden are for the most part too generalized to be of much help either in locating the specific Fox Hills outcrops that helped to form their concept of the formation or in locating exactly the vaguely defined places where they collected their fossils. Finding and re-collecting the fossil localities was particularly important for taxonomic revision and distributional studies of the Fox Hills fossils because much of the fauna described in Meek's (1876) great monograph on the paleontology of the Upper Missouri Cretaceous, including many type specimens, comes from the area of the type Fox Hills. All of these have been located with reasonable certainty by tracing Hayden's travels around the region (Fig. 3). Interestingly enough, knowledge of the details of fossil distribution in the area has proved a very useful tool in finding the localities where Hayden studied and collected during 1854 and 1855, the two years in which he accumulated most of his data on the Fox Hills. Hayden's own field notes for this period have never been found, but a study of the specimens he collected, as well as knowledge of locally abundant species that he failed to collect, have helped to clarify the scanty historical records of his wanderings. For example, his specimens reveal which of two trading posts on the Moreau River was the well known "Moreau Trading Post" locality; they also record the rather incredible fact that he never collected from the abundantly fossiliferous lower Trail City beds.

18

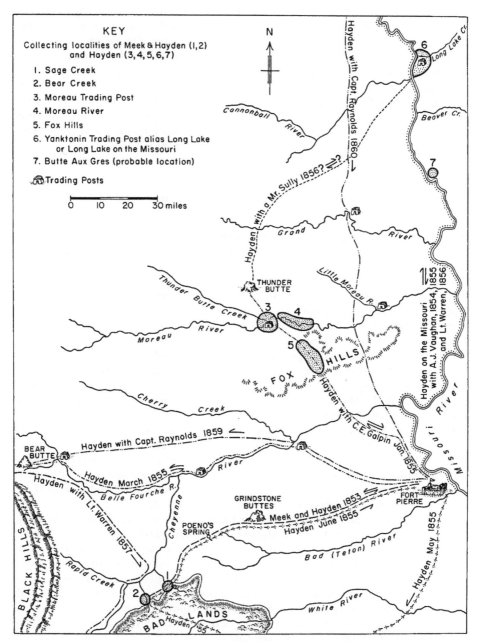

FIG. 3. Principal exploration routes and collecting localities of Meek and Hayden in the Upper Cretaceous terrains of the Dakotas (based on various sources acknowledged in text).

Contributions to the stratigraphy of the type Fox Hills since the work of Meek and Hayden all have been incidental to other objectives, either as parts of regional geologic studies and efforts to resolve the Laramie controversy, or as byproducts of areal mapping programs and surveys of water, coal and other mineral resources. This diversity of objectives was coupled with very spotty coverage of the outcrop in the type area so that workers familiar with different parts of the area, like blind men describing the elephant, formed different views as to what constituted the Fox Hills Formation.

In short, the formation has never received more than cursory study in its type area. The work of Meek and Hayden was excellent exploratory reconnaissance, but no more than that. Subsequent studies added much information and some confusion, but these were not concerned directly enough with the formation either to reveal clearly the intricacy of the stratigraphy or to attempt to interpret it.

EARLY KNOWLEDGE OF THE INTERIOR CRETACEOUS

Prior to the work of Hayden and Meek only the barest outline of the geology along the Upper Missouri was known. One of several historical summaries in their reports (Hayden, 1862, p. 1-4) credits the geographer Nicollet with "... providing the first reliable information ... respecting the extent and interest of the Cretaceous rocks in that region." In 1839, during the last of three expeditions into the Upper Mississippi region, Nicollet traveled up the Missouri as far as Fort Pierre collecting specimens and noting the succession of strata. As Hayden points out (1862, p. 3), Nicollet ". . . saw nothing of No. 5, though he obtained some of its characteristic fossils, which may have been presented to him by members of the American Fur Company." Thus Hayden explains the anomaly of typical Fox Hills fossils in the collections of an expedition that halted many miles south of the southernmost Fox Hills outcrop in the Missouri Valley and traveled northeastward into the James River valley and north to Devils Lake, well east of any known Fox Hills exposures. Nicollet (1841, p. 154; 1843, p. 36-39) described the Cretaceous beds he observed on the Missouri, listing four units and calling them, collectively, the Dixon (or Dixon's) group. In terms of the later classification the Dixon group included part of the Benton, the Niobrara, and as much as he saw of the Pierre Shale up to Fort Pierre.

Subsequently Edward Harris, who went up the Missouri with Audubon in 1843, had the following to say about the succession of strata (1845, p. 235). "Then we have Nicollet's great bed of clay, which is visible until you pass the Mandans, terminating between Beaver (= Beaver Creek, Emmons County, North Dakota) and Grand Rivers. Here commences what I have called the Yellowstone series, which probably continues to the great Falls of the Missouri. . . ." It is only of academic interest that prior to the work of Meek and Hayden two stratigraphic names had been introduced to cover parts of the Cretaceous succession of the Upper Missouri: the Dixon group and overlying Yellowstone series; the latter obviously included at its base what was later to be called the Fox Hills Formation. Neither of these two terms was ever employed by anyone except the men who introduced them.

In 1849 John Evans made a trip into the Upper Missouri country that had great potential as a source of information on the geology of the region. Evans was employed by David Dale Owen who, at the time, was engaged in his notable geological survey of Wisconsin, Iowa and Minnesota for the General Land Office of the Federal Government. Owen's objective was to take the lead in opening the rich bone beds of the White River badlands, just recently revealed (Prout 1847), still unexplored and temptingly within reach. The success of this opportunistic move presumably discouraged the Washington office from any censure of Owen for sending Evans to geologize so far beyond the prescribed limits of the survey.

Accompanying a party of the American Fur Company, Evans went to the badlands and collected from them. Subsequently he saw much of the Cretaceous west of the Missouri for he apparently was with the Company party when it traveled overland to Fort Union through the heart of what is now the type area of the Fox Hills. Owen (1852, p. 195-206) published the "substance" of Evans' report to him but aside from brief mention of Cretaceous strata it deals only with the badlands and its spectacular fauna. No complete report of Evans' travels was ever published and his opportunity to anticipate the work of Hayden and Meek went unrealized except for the description of a few invertebrate fossils by Owen (1852) and by Evans and Shumard (1854 and 1857). Evans' report to Owen apparently made no attempt to establish a stratigraphic sequence of units but the name "great lignite formation", used informally by Owen (1852, p. 195), makes its first appearance in the geologic literature of the interior region.

Evans' collections from the Fox Hills Formation include specimens from both the Trail City and Timber Lake members. Localities given with the specimens indicate that he visited both the Moreau and Grand River valleys as well as the 'Fox Hills" divide between the Cheyenne and Moreau. Although the collection he described with Shumard has been lost, the special mention of the occurrence of their bivalve species *Limopsis striatopunctatus* Evans and Shumard in great numbers establishes that Evans collected from concretions in the *Limopsis—Gervillia* Assemblage Zone of the Trail City Member, for *Limopsis* occurs only as rare scattered specimens in the formation except at this horizon. There is strong indication that this and other species from the "Grand River" were collected from the prolific locality at Bullhead, in Corson County. A member of the Evans party, de Girardin, subsequently wrote a popular account of the trip (1936) in which he mentions camping on a hill overlooking the Grand River and observes (1936, p. 71) that "The bed of this stream of water is strewn with a quantity of round stones easy to break and containing a great variety of shells and petrified snails in a perfectly preserved state." The highly fossiliferous concretions of the lower Trail City beds crop out at river level and are concentrated in the bed of the Grand only in the vicinity of Bullhead. Moreover, the selection of this campsite was probably not fortuitous as the American Fur Company maintained a trading post here as early as 1831 (Deland, 1918, p. 234).

Evans' small collection from the exceptionally fossiliferous beds of the Trail City Member is significant chiefly because Hayden appears to have overlooked these beds completely in his subsequent, more extensive collecting in the type area of the Fox Hills. Haydens' omission can in part be explained by the fact that his travels only once (see p. 26) took him through the restricted area of distribution of the fossiliferous Trail City beds, but Evans' failure to make larger collections from the rich localities through which the Fur Company party traveled suggests that he had had a surfeit of fossil collecting in the badlands.

FORMATION NO. 5 OF MEEK AND HAYDEN

THE INITIAL FIELD STUDY

In 1853 James Hall, then State Geologist of New York, sent F. V. Hayden and F. B. Meek to collect in the White River badlands. The excursion was to be Meek's only trip into the Upper Missouri country, but the first of many for Hayden. In the famous and apparently untroubled collaboration that followed, Hayden took the role of field man and collector and Meek the role of advisor and paleontologist. The joint publications that permanently linked their names began in 1856 and reached a high point in 1861 with their most detailed paper on Upper Missouri stratigraphy in which names replaced numbers on the five units of their original classification of the Cretaceous sequence. Several summaries of their conclusions were presented as their collaboration progressed and, subsequently, Hayden alone published reviews of Upper Missouri geology in 1862 and 1869, but no single publication gives a complete account of their information and interpretations. Consequently, a thorough combing of all their writings is required to bring together the scattered data and ideas that make up their interpretation of the stratigraphy. This is particularly true in attempting to understand their concept of Formation No. 5.

Meek's (1853, unpublished) journal of the badlands trip reveals that they first saw exposures of what was to become Formation No. 5 on the trail (Fig. 3) from Grindstone Butte to Poeno Spring (Meek spells this Pino's Spring; it was a popular camp site on the Fort Pierre— Fort Laramie trail located at the head of what is now Poeno Creek, a small tributary to the North Fork of the Bad River.) Between these landmarks Meek measured a section, which he replaced on the return trip with a more complete one, at a locality that cannot be exactly located from his descriptions. The following is an abridgement of this section from Meek's journal entry of July 14th, during the return trip. The beds are numbered in descending order.

No. 1. Thinly but not very distinctly laminated white limestone—about 2 feet thick at top—slope 20 feet made up with fragments of same (loose) ..... 22 feet

No. 2. Light grayish brown indurated clay .... 14 feet

No. 3. White indurated clay .... 10 feet

No. 4. Yellowish often laminated argillaceous rock, sometimes approaching a sandstone in composition, containing fucoidal marking tracks and trails resembling some of those in the Clinton group of N.Y. also a few thin seams of white limestone .... 40 feet

No. 5. Thinly laminated or foliated fine grained sandstone .... 18 inches

No. 6. Indurated clay slightly more greenish tinge than No. 4 .... 16 feet

No. 7. Rock similar in composition to No. 5 but not slaty in structure .... 12 inches

No. 8. Very hard sandy clay—light yellowish green tinge—very similar to No. 6 but harder. Containing a few large dark brownish gray nodular concretions marly, similar in composition to those in which we found the nodose baculite on Bear Creek .... 16 feet

Meek notes that units 1, 2 and 3 are probably a Tertiary outlier. Although he found no fossils in the underlying clays and sands of units 4 through 8 he equates them with

Cretaceous beds he had seen in a similar position directly beneath the Tertiary in the badlands. Here Meek had measured sections at the heads of Bull Creek, Sage Creek and Bear Creek and at all three localities had found a zone of yellow, pink and brown beds of varying lithology between the Tertiary beds and the underlying gray shales with Cretaceous fossils. In the Bear Creek section he collected "a large nodose Baculite" only 20 feet below the Tertiary contact in the varicolored beds and one of his sections on Sage Creek yielded some crushed ammonites and pelecypods from the varicolored beds. As both the varicolored zone and the sandy beds of the Bad River section were overlain by Tertiary and underlain by gray Cretaceous clays it was natural for Meek to consider them the same formation and to base his impression of Formation No. 5, the highest Cretaceous unit, on these exposures.

The first product of the badlands trip was a paper by Hall and Meek (1856) primarily devoted to the description of new species of fossils but also containing the first classification of the interior Cretaceous. The latter consists of a briefly annotated section listing the five units considered by Meek (and presumably, Hayden) to be the major subdivisions of the Cretaceous sequence as they had seen it along the Missouri Valley to Fort Pierre and across the plains to the badlands. Formation No. 5 is described as 80 feet of "Arenaceous clay passing into argillo—calcareous sandstone."

This first description can only have been based on the Bad River and badlands exposures. They saw no other exposures of No. 5, actual or presumed, for their route of travel elsewhere lay well south of its area of outcrop (Fig. 3).

Baculites grandis is the only fossil described from No. 5 by Hall and Meek (1856, p. 402); this is the "large nodose Baculite" Meek mentions in his journal as coming from 20 feet below the Tertiary in the Bear Creek section. As many geologists did after them, Meek and Hayden mistook the varicolored beds beneath the Tertiary in the badlands for a discrete stratigraphic unit and accepted it as Formation No. 5. Nearly 80 years passed before it was recognized that these varicolored beds are a complex weathered zone that developed on a number of Cretaceous formations prior to the deposition of the Tertiary beds. At the Bear Creek locality this weathered zone is in the Pierre Shale. As a result of their misunderstanding of these relationships *Baculites grandis* became one of the commonly listed guides to Formation No. 5 in all of Meek and Hayden's subsequent publications, although they never reported it from No. 5 at any other locality.

It is evident from his journal that Meek was largely responsible for the stratigraphic work on the badlands trip. Some of the impressions of No. 5 he gained from it persist throughout his and Hayden's writings.

HAYDEN'S FIELD WORK

Following the trip to the badlands for Hall, Hayden returned to Albany with Meek, but in the spring of 1854 he severed his connection with Hall and began two years of travel in the Upper Missouri country. A large part of his subsistence and travel was furnished by Col. A. J. Vaughan, Indian Agent, and he received considerable assistance from the American Fur Company with whose voyageurs and agents, particularly Alexander Culbertson and Charles Galpin, he also lived and traveled.

During this period Hayden accumulated most of his information on Formation No. 5, visiting and collecting from the type area and neighboring outcrop areas bordering the Missouri River in what is now North and South Dakota. Lacking his field journals, only a sketchy outline of the route of his travels can be pieced together from the fragmentary references in his publications and letters and from the evidence of his collections. In his own terse summary of these wanderings, he '. . . explored the Missouri to the vicinity of Fort Benton, and the Yellowstone to the mouth of Big Horn river, also considerable portions of the Bad Lands of White river and other districts not immediately bordering on the Missouri" (Hayden, 1862, p. 4.)

From a few indirect references in his papers and from some of his letters to Joseph Leidy, we know that he started up the Missouri in the spring of 1854 and, during that summer, followed it to the Yellowstone, which he ascended at least as far as the mouth of the Big Horn River. In November he returned to Fort Pierre where he spent a busy winter making excursions into the surrounding country. In a letter to Spencer F. Baird, of the Smithsonian Institution, from Fort Pierre dated Feb. 9th, 1855, Hayden wrote: "I have just returned from the Moreau trading post. 14 days travel from Fort Pierre. I had Mr. Gilpin as companion who feels much interest in Geology and has a pretty good practical knowledge of it. I had fine opportunity for examining the Fox Hills, Moreau River, and many other places of much interest and made a fine collection of Shells and large bones."

Whether Hayden made more than one trip with Charles Galpin to the Moreau River that winter is not known; after the trip mentioned in his letter to Baird, most of his time seems to have been occupied in trying to return to the badlands. On the same day (Feb. 9, 1855) that he wrote to Baird, he sent a letter to Leidy in which he noted that he was about to leave for another collecting trip, sponsored by Col. Vaughan, to the badlands. This trip was stopped at Sage Creek, short of its goal, by a blizzard (Hayden 1857b, p. 152), but apparently this did not deter Hayden from an exploratory side trip, for we have his own statement (Hayden, 1856, p. 71) that on March 9, 1855, he climbed Bear Butte, about 65 miles airline northwest of Sage Creek, and well off the trail back to Fort Pierre. Hayden succeeded in reaching the badlands later that spring, leaving Fort Pierre on May 7 and returning to it June 6. An account of this trip, drawn from one of the long-missing Hayden journals, is included by Hayden in his first publication (Hayden 1856, p. 71-76).

Several weeks after his June 7th return from the badlands he left Fort Pierre for Fort Union and a summer on the upper reaches of the Missouri and its tributaries. From subsequent letters, in particular one to Baird from St. Joseph, Missouri, dated December 20, 1855, we know that Hayden traveled by steamboat from Fort Pierre to Fort Union, thence by keelboat to Fort Benton. As was his custom he undoubtedly made studies en route and his collections from this season indicate he was also on the Milk River, but his most intensive geologic work during the summer of 1855 was on the Judith River badlands of the Upper Missouri. He notes in the letter to Baird that he "started from the mouth of Judith 22nd October, froze up at Fort Pierre, and then came the rest of the distance by land to this point." Later in the winter he returned to Washington and then in February of 1856 joined Meek in Albany to spend the spring working up some of the invertebrate fossils from his collection. Hayden's letter of transmittal accompanying Meek's monumental paleontologic monograph of the Upper Missouri fossils (Meek, 1876, p. 111) clearly states that "The accumulation of the materials which compose this volume was commenced in the spring of 1854, and the greater number of the new species of fossils were discovered by the writer of this letter during that and the succeeding year." This is borne out by the locality data supplied with the initial description of many of these fossils that appeared in papers by Meek and Hayden in volume 8 of the *Proceedings of the Philadelphia Academy of Natural Sciences*, 1856. Not only was the bulk of the Fox Hills fauna described in these papers, but they also contain most of Hayden's observations on the stratigraphy of No. 5, to which his subsequent work in the Upper Missouri country added little that was new. That four of the papers by Meek and Hayden (1856a, b, c, d) in volume 8 of the *Proceedings* were all based on Hayden's explorations during 1854 and 1855 is a matter of record. The last of these was submitted for publication June 10, 1856, as noted in minutes of the meeting of that date in the *Proceedings*, v. 8, p. 105, before Hayden had left Fort Pierre to spend the summer season of 1856 in the field as physician and geologist with an exploring party of the Corps of Topographical Engineers, U.S. Army, under Lieut. G. K. Warren. The fifth paper in volume 8 (Meek and Hayden, 1856e), published in November, is a significant paper on Upper Missouri geology largely because it contains the first catalogue of Upper Missouri Cretaceous and Tertiary fossils as well as considerable discussion and a revised section—their second. Although based largely on Hayden's field work in 1854 and 1855, a part of it derives from his trip up the Missouri with Warren in the summer of 1856. Hayden included credit to Warren's expedition in the provisional title when the paper was submitted for publication (*Proc.* v. 8, p. 260), but any mention of the expedition was omitted when the paper was published and other than the vague statement that ". . . some additional collections have come to hand . . ." (p. 265) the paper appears to stem entirely from Hayden's field efforts of 1854-55. Possibly the credit was withdrawn because so little of the new material was collected on the Warren expedition; most of the new fossils described are from Hayden's collections on the Milk and Judith Rivers, which the Warren expedition did not reach. The sole new species listed from Formation No. 5, however, is *Mactra* (*Cymbophora*) *warrenana* Meek and Hayden named "in honor of Lieut. G. K. Warren, of the U.S. Topographical Engineers" (p. 271) and collected at one of the expedition's stops along the Missouri south of what is now Bismarck, North Dakota. Warren's pique at Hayden's failure to acknowledge the role of the Corps of Topographic Engineers apparently was not lessened by the honor of having his name attached to a fossil clam, and the incident came close to terminating Hayden's valuable association with Warren's party and subsequent C.T.E. expeditions.

During the field season of 1856 Hayden, with the Warren party, went up the Missouri by steamboat to Fort Union, explored the lower part of the Yellowstone then came slowly back down the Missouri examining the country along its banks. This return trip furnished Hayden with some additional localities and observations on No. 5, particularly on its upper contact with the Great Lignite Group. The principal geologic contributions are contained in two papers (Hayden, 1857a; Meek and Hayden, 1857) that together form the core of Hayden's stratigraphic work on the Missouri Valley section. They include his first geologic map of the Upper Missouri, the first expanded discussion of the units of the Cretaceous and Tertiary section, and additional details of the Great Lignite Group.

In 1857 Hayden was with Warren on another C.T.E. expedition, this time to the

Black Hills by way of Fort Laramie and back down the Niobrara River, and in 1858 he was in Kansas with Meek; neither trip touched the Upper Missouri terrains of No. 5. Hayden's next journey, with Raynolds' Yellowstone Expedition in 1859 and 1860, also took him far afield from the Cretaceous of the Dakotas. Raynolds' return trip, however, passed directly through the type area of the Fox Hills between August 29, 1860, when they camped on the Cannonball River near the present site of Breien, and September 5 when they crossed the divide between the Moreau and Cheyenne and reached the latter river. Except for Sunday, September 2, it was apparently a steady march with little time for collecting. Raynolds invariably observed Sunday in camp, and this day the camp was located on High Bank Creek about six miles southwest of its confluence with the Grand River—in the heart of abundantly fossiliferous Fox Hills beds. Raynolds' journal (1859-60, unpublished) makes no mention of Hayden's activities during this return trip and Hayden's detailed report on the Raynolds' expedition (Hayden 1869) lacks any geological observations after leaving Fort Union.

No new fossils were reported from the Fox Hills of the Missouri Valley region in the preliminary report on the paleontology of the Raynolds Expedition (Meek and Hayden, 1861). It is in this report, however, that the Cretaceous units are given their most detailed descriptions and assigned names. Moreover, it contains the authors' most exhaustive treatment of the correlation of their interior Cretaceous section with those elsewhere in North America and in Europe. Of all the papers by Meek and Hayden this is probably the most informative on the subject of the Upper Missouri Cretaceous rocks. It contains more stratigraphic data on this subject than Hayden's general summary of Upper Missouri geology (Hayden, 1862) and gives slightly more detail than the introductory statement of Meek's Upper Missouri paleontology (Meek, 1876, p. XIX-LXIV), although the latter has a more complete analysis of the paleontology and its implications. The 1861 paper draws heavily from preceding papers by the authors and can be considered as their final summary on the Cretaceous rocks of the Dakotas; all subsequent papers by either author appear to derive their information on the subject directly from this report, much of Hayden's summaries of 1862 and 1869 being verbatim from it.

Hayden's direct contact with Formation No. 5 in the Missouri Valley region of the Dakotas came to an end with the return trip of the Raynolds Expedition in 1860. Indeed, any fruitful observations on, or collections from, Formation No. 5 in this area must have ended with the Warren expedition of 1856. No new material appears in Hayden's writings subsequent to the report on this expedition (Meek and Hayden, 1857). After the Civil War, Hayden went collecting in the badlands for the Philadelphia Academy in 1866 and made a geological survey of the State of Nebraska in 1867; subsequently all of his work was in the Rocky Mountain States in connection with his U.S. Geographical and Geological Survey of the Territories.

THE INTERPRETATION OF FORMATION NO. 5

Early in their work on the Cretaceous-Great Lignite section in the Missouri Valley, Meek and Hayden recognized that they were dealing with a relatively continuous succession of beds and they emphasized the gradational nature of their subdivisions and the continuity of the whole in a number of their papers (Meek and Hayden, 1856d, p. 113; 1861, p. 432; Hayden, 1856, p. 156; 1862, p. 128-129). To the upper part of the section they paid particular attention, demonstrating the gradual change from marine to fresh-water conditions by careful documentation of the successive changes in faunas and sediments. Their criteria for the recognition of individual units in the succession are 1) lithology and 2) diagnoses of both age and environment based on the faunal content. For most of the Cretaceous units—Formations No. 1 through No. 4 (Dakota, Benton, Niobrara, Pierre)—lithology is the primary diagnostic feature, but Formation No. 5 has its lower boundary defined by lithology and its upper boundary defined by age and genesis. In terms of the present definition of the Fox Hills Formation, Meek and Hayden's basal contact is too generalized to permit specific identification at any given place. Their upper contact, however, was drawn at the base of the sands of the Colgate lithofacies in the top of the present Fox Hills (Iron Lightning Member).

The Lower Contact and Faunal "Blending"

The base of Meek and Hayden's Fox Hills is defined on lithologic grounds in their 1861 paper (p. 427) where they state:

> This formation is generally more arenaceous than the Fort Pierre group, and also differs in presenting a more yellowish or ferruginous tinge. Towards the base it consists of sandy clays, but as we ascend to the higher beds, we find the arenaceous matter increasing, so that at some places the whole passes into a sandstone. It is not separated by any strongly defined line of demarcation from the formation below, the change from the fine clays of the latter to the more sandy material above, being usually very gradual.

The fact that sandy clays transitional between the clays of the Fort Pierre and overlaying sands were originally included in the Fox Hills could not be stated more clearly. But there is no record indicating that Hayden actually considered the clayey silts (Trail City Member) of the type area a part of the Fox Hills. In the earliest sections published, the Fox Hills is described (Hall and Meek, 1856; Meek and Hayden, 1956a, p. 63; 1856e, p. 269) as "Gray and yellowish arenaceous clays . . ." and no mention is made of sand or sandstones. This undoubtedly stems from Meek's confusion, previously noted, of the varicolored weathered zone at the top of the Pierre with Formation No. 5 in his Sage Creek and Bear Creek localities of the badlands.

It was this same confusion with weathered Pierre Shale that led Meek to list *Baculites grandis* Hall and Meek as a guide fossil to Formation No. 5. The faunal relationships of the Pierre and Fox Hills were further obscured when *Baculites ovatus* Say and *Ammonites [Placenticeras] placenta* Dekay were listed as common to both these formations (Meek and Hayden, 1857, p. 128); neither species occurs in the Fox Hills Formation of the type area or adjacent regions. Most likely Hayden found the species in weathered Pierre Shale of the badlands region during one of his later trips to that area (Fig. 3). Hayden was perplexed by the badlands stratigraphy for a number of years, believing that the White River beds were a lateral phase of his Great Lignite Group; consequently he assumed that the White River beds, like the Lignite Group, were gradational downward without break into Formation No. 5 of the normal Cretaceous sequence. Ascribing the weathered zone beneath the irregular unconformity at the base of the White River to the Fox Hills (Hayden 1858, p. 156) helped to preserve this illusion. Even when Hayden finally was able to demonstrate a marked difference in age between the White River beds and the Great Lignite Group, neither he nor Meek followed the implications of this change in interpretation far enough to make the necessary corrections in their faunal lists.

These misunderstandings regarding the distribution of faunas led Meek and Hayden (1861, p. 427) to append the following somewhat ambiguous statement to their lucid lithologic description of the Pierre-Fox Hills contact:

> Nor are these two formations distinguished by any abrupt change in the organic remains, since several of the fossils occurring in the upper beds of the Fort Pierre Group pass up into the Fox Hills Beds, while at some localities we find a complete mingling in the same bed of the forms usually found at these two horizons. Indeed, we might with almost equal propriety, on paleontological principles, carry the line separating these two formations down so as to include the upper fossiliferous zone of the Fort Pierre Group, as we have defined it, in the formation above.

In several places in their works Meek and Hayden refer to this mixture of Pierre and Fox Hills species; as Hayden states it (1862, p. 81), "At Sage Creek and on the Yellowstone . . . the fossils indicate a blending of Nos. 4 and 5." The Sage Creek here referred to is the one flowing northwestward into the Cheyenne River southwest of the town of Wall, South Dakota. Heading in the White River beds, its main valley exposes highly fossiliferous Pierre Shale, the site of the famous Sage Creek collecting locality (Fig. 3, locality 1). Here, Pierre fossils in the weathered zone beneath the White River beds were undoubtedly attributed to the Fox Hills, as was *Baculites grandis* on Bear Creek, a few miles to the southwest. The Yellowstone River occurrence mentioned by Hayden is apparently based on his first encounter with the Cretaceous on the Yellowstone in the vicinity of what is now Glendive, Montana, at the north end of the Cedar Creek anticline. The impression that the Fox Hills and Pierre faunas were "blended" in this area was probably gained by using the incorrectly placed Bear Creek and Sage Creek fossils as a basis for comparison.

In the end, Meek and Hayden put more store by their lithologic criterion and were not sufficiently influenced by their interpretation of faunal "blending" to change the basal contact of No. 5, for they add the following to their discussion of the mixed faunas (1861, p. 427): "All the facts, however, so far as our present information goes,—taking into consideration the change in the sediments at or near where we have placed the line between these two rocks,—seem to mark this as about the horizon where we find evidences of the greatest break in the continuity of physical conditions."

Meek's (Meek and Hayden, 1856e, p. 266) early recognition of major faunal groupings in the Cretaceous section was a perceptive paleontologic generalization. His grouping of Formations 4 and 5 as a gross faunal unit is essentially correct, for during the time of Pierre-Fox Hills deposition the endemic fauna of the Late Cretaceous interior seaway apparently evolved without any cataclysmic changes, though it was subject to the periodic immigration of new elements from both the boreal and Gulf regions. Nevertheless, the Fox Hills fauna in the Dakotas is easily distinguished from any of the faunas in the underlying Pierre shale by its ammonoids. Meek and Hayden's inclusion of Pierre ammonoids of the badlands in the Fox Hills fauna prevented their detecting this. Hayden (1862, p. 80) wrote in a summary of the Fox Hills faunas that: "The greatest proportion of the species are restricted to this bed; and those which are common to it and formation No. 4 are chiefly *Cephalopoda,* which everywhere have an extensive vertical as well as geographical range." The reverse of this statement is much closer to the truth; the Pierre and Fox Hills of the Dakotas have many gastropod and pelecypod species in common but, except for a single species each of a nautiloid and belemnoid, the abundant cephalopod faunas are quite distinctive.

TRANSITIONAL UPPER CONTACT

The gradual transition from marine to nonmarine beds at the top of their Cretaceous sequence offered Meek and Hayden some of their most challenging stratigraphic and paleontologic problems; as a consequence, their discussions of these beds cover many more pages in their writings than do their descriptions of the relatively simple marine Cretaceous sequence beneath. Their first view of Cretaceous beds in contact with Tertiary beds in the White River badlands, where marine ammonoid-bearing beds give way abruptly to freshwater mammal-bearing beds, was so clear-cut it permitted but one solution—the top of the marine beds (mistakenly thought to be Formation No. 5) was the Cretaceous-Tertiary boundary.

In the Missouri Valley, where Hayden subsequently found lithologic gradation and an accompanying transition from marine through brackish-water to fresh-water faunas, the problem was much more complex. Here they chose, as the basal bed of their Tertiary, the first "estuarine" bed marked by a brackish fauna containing, among other species, *Crassostrea, Melania,* and *Corbicula,* but lacking, in their collections, exclusively marine invertebrates such as ammonoids. This bed, which they refer to as Bed Q in the section of their Great Lignite Group (Meek and Hayden, 1857, p. 122), is described as "Gray compact, or somewhat friable concretionary Sandstone" overlain by lignite. From numerous references to Bed Q at a number of localities, including some in the Moreau and Grand River valleys there can be no doubt that it includes widespread, though lenticular, sands of the Colgate lithofacies that occur at the top of the Fox Hills as presently classified (see Fig. 4).

It is explicit in the writings of Meek and Hayden that the choice of the base of Bed Q as the Cretaceous-Tertiary boundary depended on their interpretation of the invertebrate faunas. This decision did not come easily for they were impressed by what they considered the strong Tertiary affinities of many elements in the fauna of Formation No. 5. In one of several passages discussing Bed Q which express their quandary they stated (1857, p. 122-123):

Near Long Lake—and on the Moreau, the entire bed [Bed Q] is exposed, and attains a thickness of about thirty feet. At these latter localities it is seen to repose directly upon No. 5 . . . the upper part of which it so nearly resembles in its lithologic characters that the line of demarkation between the two can often be only ascertained by the organic remains characterizing each. This fact, together with the general resemblance of many of the fossils found in the upper part of No. 5 . . . to Tertiary types, would have caused us to doubt the propriety of referring this part of that bed (No. 5) to the Cretaceous epoch, were it not for the presence of *Scaphites conradi*, and other well marked Cretaceous forms.

Again, in Hayden's summary of the Upper Missouri geology (1862, p. 128) we find the following statement in his argument for placing the Lignite Group in the Tertiary:

> We have also mentioned the fact that the fossils of the upper part of No. 5 seem to have existed upon the verge of the Tertiary period, that they sometimes present peculiar forms more closely allied to Tertiary types than Cretaceous, and were it not for the presence of the genera *Baculites, Ammonites, Inoceramus,* etc., which are everywhere supposed to have become extinct at the close of the Cretaceous epoch, we would be in doubt whether to pronounce them [the fossils of No. 5] Tertiary or Cretaceous.

At least in the early stages of work (Meek and Hayden, 1856e, p. 268) Meek was convinced that the fauna of Bed Q was Tertiary both by his favorable comparisons of the brackish- and fresh-water invertebrates with foreign faunas of Miocene and Eocene age, and also by the affinities of his fossils with modern species. Moreover, both men at first were persuaded by Newberry that the fossil floras of the Great Lignite Group were Miocene.

Obviously ammonoids were a critical factor in separating Bed Q from the marine beds into which it graded below. One wonders whether Hayden would have included Bed Q in Formation No. 5 had he come across some of the ammonoids that were subsequently found in it; if so, the upper contact of Formation No. 5 would have corresponded with that of the present Fox Hills Formation.

Vertebrate remains did not play a significant part in this early dating of the Missouri Valley units. In hindsight it may seem strange that the dinosaur bones Hayden saw in the Dakotas did not make more of an impression on him but at that time Hayden still believed the Great Lignite and White River beds were correlative facies and he assumed all large bones were those of brontotheres. The earliest mention of large bones from the Missouri Valley area (Meek and Hayden, 1856d. p. 116) concerns specimens from what I judge must be one of the channel deposits of Colgate lithofacies near the confluence of Thunder Butte Creek and the Moreau River. In a later paper this occurrence is used (Hayden, 1857a, p. 120) to support the correlation of the Great Lignite and the White River beds:

> Among other facts . . . we would mention that a friable sandstone seen crowning some of the hills near Moreau River, in which specimens of a *Cyrena* [= *Corbicula*] . . . and *Ostrea* [= *Crassostrea*] *subtrigonalis* (E & S) were found associated with large bones supposed to be those of *Titanotherium,* and which bed we had regarded as probably a distant outlier of the White River formations, was found to be the same as the lowest bed of the Great Lignite basin. . . .

Interestingly, a footnote on p. 121 of this same paper noted that Leidy, who first referred these bones to *Titanotherium?*, now suspected that they belonged to a "huge

Dinosaurian" he had previously described (Leidy, 1856, p. 311) as *Thespesius* (a hadrosaurid). So Hayden was aware, by 1857, that he had dinosaurs in his "Bed Q". We also know that on his side trip to Thunder Butte during Warren's 1856 expedition he saw (Hayden, 1862, p. 100) "bones belonging to some huge sauroid or manatoid animal . . . scattered very abundantly over the country. . . ." However, he obviously was not completely certain that these were dinosaur bones nor, apparently, was he aware of the restriction of dinosaurs to Mesozoic beds at that time.

One other subject relating to the upper contact of the Fox Hills Formation as interpreted by Meek and Hayden concerns the nature of "Bed Q," the limiting bed marking the base of their Lignite Group. As was noted in the stratigraphic summary, Colgate-like lithology, although concentrated in the uppermost part of the formation, appears locally at a number of levels in the "banded-beds" of the Bullhead lithofacies and in the lower part of the Hell Creek Formation (Fig. 5). From their writings it is apparent that Meek and Hayden believed Bed Q to be a single widespread bed, but they were only slightly more fallible in picking out this unit at different localities than later geologists have been in trying to use the Colgate lithofacies as a mappable unit. Meek and Hayden knew nothing of the marine Cannonball Formation and where they encountered these beds they consistently referred them to their Formation No. 5, which it closely resembles lithologically, and picked the adjacent overlying brackish or freshwater sandy beds in the Fort Union as their Bed Q.

To summarize, Meek and Hayden considered the upper boundary of Formation No. 5 to be the horizon at which the faunas of the beds in the gradational sequence changed from marine to brackish-water species. They believed this to be a widespread, abrupt and contemporaneous faunal change marked by Bed Q and its "estuarine" fauna, but they specifically note that Bed Q is gradational beneath with the upper beds of Formation No. 5. They considered that this same environmental change marks the passage from the Cretaceous to the Tertiary period in the Upper Missouri Valley.

FAUNAS, AGE AND CORRELATION

The pioneer studies of the fossils from the interior Cretaceous were largely the work of Meek. Hayden spent little time assisting in the paleontologic chores except for his brief return to Albany in 1854 to work on collections with Meek, and it is obvious that the sound generalizations about the faunas could only have come from a student with Meek's background.

A number of Meek's broad observations have become basic ingredients of our present-day understanding of the interior faunas. He singled out the major faunal breaks in the Cretaceous sequence by recognizing three gross faunal groupings (Meek and Hayden, 1856e, p. 266): the fauna of Formation No. 1 (Dakota), those of Formations No. 2 and 3 (Benton and Niobrara), and those of Formations No. 4 and 5 (Pierre and Fox Hills); and he pointed out how markedly distinct these three groups were from one another. As a consequence of this observation the authors were able to match the interior Cretaceous with the standard European sequence of the time with a fair degree of accuracy.

In the same paper (p. 265-266) Meek and Hayden draw attention to the overall peculiarity of the interior Cretaceous marine faunas, pointing out the great preponderance of molluscs and the relative scarcity of most other common marine inverte-brate fossils. This empirical fact has been "rediscovered" a number of times since but still lacks a satisfactory explanation.

Meek and Hayden correlated the Pierre-Fox Hills faunas with the uppermost beds of the New Jersey, Mississippi, and Alabama Cretaceous, chiefly on the basis of cephalopods. The details, summarized in Meek, 1876, are unimportant here. Subse-quent work corrected certain errors in taxonomy and faunal distribution and fur-nished more stratigraphic detail as the geology of all these areas became better known. Scrutiny of the faunal lists given by Meek and Hayden reveals that they were prevented from making more exact comparisons of the Fox Hills fauna because of their inclusion in it of elements of the Sage Creek fauna of the Pierre Shale in the White River badlands. Even so, the error made little difference at the time, for, with the possible exception of the Alabama sequence, the stratigraphic distribution of Late Cretaceous ammonoids was no better known in the Coastal Plain, where they are relatively scarce, than in the Missouri Valley area.

In relating the Pierre-Fox Hills faunas to the European Cretaceous sequence Meek and Hayden (1861, p. 428-432) present a thorough analysis of the intercontinental affinities of the American Late Cretaceous faunas in general. Interestingly, no ammo-noids are used in this comparison. Meek deduced correctly that the Pierre-Fox Hills faunas (p. 432) ". . . belong to the horizon of the Upper or White Chalk and Maestricht Beds of Europe"—Campanian and Maestrichtian of present usage.

The great geographic and stratigraphic sweep of Hayden's field work on the Upper Missouri geology did not permit much attention to detail. Perhaps his experience led him to rely on the marine units being considerably more persistent and uniform than the nonmarine units, and accordingly concentrated on the more apparent puzzles of the White River beds and Great Lignite Group. On the other hand, it is more probable that Meek's grasp of the marine stratigraphy from his one encounter with it dominated their collaboration. It is a fact that the first section of the marine units (Hall and Meek, 1856, p. 405) was drawn up by Meek and that it changed only in details during subsequent revisions. Hayden's chief contribution to the marine section thereafter was the data on Formation No. 5 that he accumulated with Galpin in the type area during the early months of 1855 (Fig. 3). Whatever the reason, Hayden did not pay very close attention to the marine beds under the Great Lignite after 1855. Although he collected some of the critical evidence for it, Hayden's field work on the Upper Missouri Cretaceous understandably was not detailed enough to permit the two men to recognize that Formation No. 5 markedly transgressed time.

The same combination of broad reconnaissance and reliance on the persistence of the established marine sequence undoubtedly contributed to their miscorrelation of the Judith River beds and to mistaking the Cannonball beds for the Fox Hills. Stanton (1920, p. 1 and 11) drew attention to the inclusion of the five species of Cannonball fossils in the Fox Hills fauna by Meek and Hayden; as already noted, this was a case of mistaken identity of similar lithologies that could only have been avoided by more detailed field work. The problem of the Judith River beds was more complex. Meek and Hayden (see Meek's summary 1876, p. XLVII-L) first correlated these brackish- and fresh-water beds, found at the confluence of the Judith and Missouri Rivers, with their Formation No. 1 primarily because Leidy thought the vertebrate remains resembled those from the English Wealden (Lower Cretaceous), and also because the nonmarine invertebrates were different and "older looking" than Fort Union species, although undiagnostic of any other horizon. In addition they described (1856e) ten species of marine invertebrates from sandy beds at the base of the Judith River succession which they also attributed to Formation No. 1. Subsequently, additional occurrences of these marine fossils led Meek (1876, p. XXXVI) to observe that they ". . . have been elsewhere discovered connected in such a way with the Fox Hills beds, and containing so many of the common fossils of the same, as to show that they form an upper member of that group." Accordingly the Judith River beds were assigned to a new position in the Cretaceous between the Fox Hills Formation and the Great Lignite or Fort Union Group, and the ten marine species were included in the Fox Hills fauna (Meek, 1876, pls. 37, 38, 39). As is well known, later studies of the Judith River beds have revealed that they are a complex wedge of brackish-water and fresh-water sediments that thins out eastward within the Pierre Shale equivalent, the faunal resemblance of the Judith River marine species to Fox Hills species being due to similarity in environment, not to similarity in age.

COLLECTING LOCALITIES

The knowledge that Hayden made at least three trips into or across the type area of the Fox Hills (see Fig. 3) tells us little about what he actually saw of the formation or where he collected from it. His publications contain no specific references to stratigraphic features in the type area below his ' Bed Q'. A letter Hayden wrote to Baird from Fort Pierre, Feb. 9, 1855, indicates that Hayden had a great deal of information that was never published; after a paragraph about his trip to Moreau trading post with Galpin (see p. 24) he added: "I have just copied my geologic sections and they fill over 100 pages of letter paper closely written." None of Hayden's sections have ever come to light. The most helpful pieces of information in locating his collecting localities are his fossil collections themselves, the little sketch map he made for Warren, and Warren's early map of the Upper Missouri country (Warren, 1859). Also helpful was knowledge of the location of trading posts of the American Fur Company that were operative at the time Hayden was wandering about the area; this was gleaned from several sources (Warren, 1856 and 1859; Deland, 1918; and various Meek and Hayden publications). All the pertinent trading posts are shown in Fig. 3.

An examination of Hayden's collections from the type Fox Hills and adjacent areas, all the data on which are in Meek's (1876) monograph, reveals some odd omissions. Conspicuously absent are about half a dozen species of scaphitids that are very common in the Trail City Member over a limited part of the eastern side of the type area (Figs. 16 to 18). Species found in this restricted area include *Discoscaphites roanensis* Stephenson, *D. conradi* (Morton) *s.s.*, and several undescribed forms—a rich and highly distinctive assemblage in which many species probably represent a locally successful colonization by immigrants from the Gulf Coast during the deposition of a part of the Trail City Member. Not only are these locally abundant ammonoids missing from Hayden's collections, but three abundant bivalves that occur with them, *Limopsis, Gervillia* and *Nemodon,* were not collected in the type area, though Hayden did bring back specimens of them from Montana and Colorado. *Limopsis* and *Gervillia* in particular are incredibly abundant in the Moreau and Grand River valleys in the eastern part of the type area. Evans, as previously noted (p. 21), collected *Limopsis* and *Nemodon* around Bullhead on the Grand River (marked by the trading post in Fig. 3). Nowhere in their writings do Meek and Hayden allude to the unusually prolific assemblages of the Trail City Member that contain these fossils, whereas even in their brief systematic note Evans and Shumard 1857) remark on the unusual numbers of *Limopsis*. Hayden would certainly have collected specimens and remarked on these occurrences had he seen them.

The richly fossiliferous beds of the Trail City Member (described in a later section) crop out in the area of the trading post at the confluence of the Moreau and Little Moreau Rivers; if Hayden had visited this post he could not have helped but find the fossils. These fossiliferous beds do not extend as far west along the Moreau as the trading post at the mouth of Thunder Butte Creek. That the latter was the "Moreau Trading Post" locality is further demonstrated by Hayden's sketch map which shows a trading post here that he obviously visited, undoubtedly on his trip with Galpin in January 1855. From this post he explored the surrounding Moreau Valley and collected the *Corbicula* specimens referred to in the quote on page 30.

Corbicula can only be found in the Moreau Valley in and west of the area shown as the "Moreau River" locality on the exploration map, Fig. 3. There is no doubt whatever that this latter locality and that of the Moreau Trading Post are approximately correct in their location, for it is the only part of the entire Moreau Valley where the combination of marine and brackish Fox Hills fossils described from these localities occur in the same stratigraphic section.

On Hayden's second trip into the type area—a side trip to Thunder Butte from Warren's expedition of 1856—his route apparently lay north and west of the exposures of the highly fossiliferous Trail City beds along the Grand River east of Bullhead. But his return march on Raynolds' Yellowstone expedition in 1860 crossed the Grand in the vicinity of the present Indian village of Bullhead, encamped a full day on High Bank Creek near the Grand River and crossed the Moreau just west of its confluence with the Little Moreau—in all three places the Trail City beds are richly fossiliferous. The reason for Hayden's seeming indifference to the local geology on this return trip can only be guessed, but it is a fact that he brought back no collections from this part of the Fox Hills Formation in the Dakotas.

One other outstanding omission from the Meek and Hayden collections of the Fox Hills fauna in the Dakotas is the genus *Tancredia*. Although they describe it from the Judith River beds—and on this basis record it as a part of their upper Fox Hills fauna—there is no record in their papers or in Meek's (1876) monograph of its occurrence in any of their localities from the Dakotas. *Tancredia* is so common in the Timber Lake Member of the Fox Hills along and north of the Grand River valley— where it is associated with the burrow *Ophiomorpha* in a well-defined biofacies—that lack of reference to either of these fossils indicates a general lack of familiarity with the upper Fox Hills throughout the northern part of its outcrop in the Missouri Valley region of the central Dakotas.

These are the more conspicuous omissions in the Meek and Hayden lists and collections; to them other similar but less striking examples could be added. It is obvious from all of these that Hayden's collections from the type area came chiefly from the Timber Lake Member and the Colgate lithofacies of the Iron Lightning

Member of the Fox Hills. This is also apparent from the lithology and preservation of the specimens themselves. Moreover the faunal associations in some of the larger specimens are distinctive enough to recognize which specific assemblage zone the specimens came from, and a few associations are even diagnostic of geographic location.

Within the areas of the localities shown in Fig. 3, all the specimens collected by Meek and/or Hayden from their localities of the same name could be duplicated. Only one of these, "Butte aux Gres", lacks the additional evidence from maps, literature, unpublished records or unique fossil distribution to substantiate the location. "Butte aux Gres" was never described or indicated on a map by Hayden. Warren used the same name for Grindstone Buttes, but this is an impossibility as Hayden indicated clearly in the following description of Formation No. 5 that "Butte aux Gres" is on the Missouri (Hayden 1857a, p. 109):

> In ascending the Missouri river it [Formation No. 5] first makes its appearance near the mouth of Grand river, about one hundred and fifty miles above Fort Pierre. Near *Butte aux Gres* it becomes quite conspicuous, acquiring a thickness of eighty or one hundred feet, and containing great quantities of organic remains. Here it forms an extension of what is called Fox Ridge, a series of high hills having a northeast and southwest course, crossing the Missouri river into Minnesota at this point.

Bluffs where the outcrop of the Timber Lake Member crosses the Missouri are most conspicuous on the east bank, and very fossiliferous as well, but it is not certainly known on which side of the river the bluffs or surrounding hills were called "Butte aux Gres."

The locality called Long Lake, or Long Lake on the Missouri, has nothing to do with the present lake by that name southeast of Bismarck, North Dakota. The little stream shown at locality 6 in Fig. 3 was called Long Lake Creek in Hayden's day (see Warren 1859, on map); the Yanktonin Trading Post was near its mouth (also Warren's map). When the bivalve *Mactra warrenana* was first described (Meek and Hayden, 1856e) its locality was given as "Yanktonin trading post"; when it was redescribed by Meek in 1876 the locality given is "Long Lake". That the two localities are simply different names for the same locality is substantiated by a specimen label in Hayden's handwriting which bears both names. This specimen, in Yale's Peabody Museum, is part of a small, representative suite of Upper Missouri Cretaceous specimens given by Hayden to J. D. Dana for the Yale collections.

Locality 5, "Fox Hills", includes the historical type locality and is extended along the path indicated by Hayden's sketch map for the reason that the Timber Lake Member is at the surface here and fossils weather out in abundance. Any of the higher parts of the divide along which Hayden and Galpin traveled, as indicated by the sketch, could have yielded the fossils from the Timber Lake Member labelled "Fox Hills" in the Hayden Collection.

EVALUATION

The significant contributions to Upper Missouri geology in the early works of Meek and Hayden are the enduring synthesis of Cretaceous and Tertiary stratigraphy and the perceptive generalizations about the interior Cretaceous sea and its faunas. Their work, outstanding exploration reconnaissance for its time, obviously cannot be judged in the context of detailed stratigraphic work. For the Fox Hills Formation, as for the other stratigraphic units involved, it is the original source of information furnishing name, gross lithology, approximate stratigraphic position, and general area of typical outcrop. Beyond this, their work lacked the stratigraphic detail and substance, and contained too many omissions and misinterpretations, to serve as an authoritative source on Fox Hills stratigraphy. Consequently its priority imposes only the broadest restrictions on the content and boundaries of the stratigraphic units it describes.

STUDIES IN THE TYPE AREA AFTER HAYDEN

The forefront of geological exploration passed westward from the Dakotas in the late 1850's, early in Hayden's ascendancy as a leading figure in western geology, leaving the area between the Cheyenne and Cannonball Rivers uninvestigated except for his few observations. Even the geography of the area was but superficially known, for Warren, whose maps unveiled so much of the Upper Missouri country, admitted (1859, p. 39) that he had never been beyond the mouths of the Moreau, Grand, and Cannonball Rivers and had to rely on information from Indians and trappers in plotting their courses. The area remained an island of relatively unknown territory through the Civil War years, bypassed by the major immigration routes and military roads to the western territories that led along the Missouri to the north and along the Cheyenne, Bad, and Platte Rivers to the south. After the war this isolation was perpetuated by the treaty of 1868, which set aside for Indian occupancy all of what is now South Dakota west of the Missouri. Throughout the Indian Wars and treaty "adjustments" of the 1870's, this country became particularly inhospitable and during the progressive restriction of Indian lands that followed, it escaped white settlement until 1909.

During this nearly half century of isolation the geology of the region received scant attention. In contrast, other parts of the western interior were extensively explored and studied. The lithogenetic equivalents of the Fox Hills Formation were observed flanking the ranges and lesser uplifts in much of the northern interior region, and the name Fox Hills Sandstone was commonly applied to them in parts of Colorado, Wyoming, and Montana. Substitution of the word *sandstone* for *formation* in these areas involved a change in the interpretation of the lower contact of the formation: the sandy clays transitional with the underlying Pierre Shale were included with the Pierre, and the Fox Hills contact became the base of the first sandstone in the sequence. Probably the chief reason for this change was that the base of the sandstone furnished an obvious topographic break that facilitated mapping. The sand content and grain size of the Fox Hills increases west of the type area and in eastern Wyoming the sandy beds are noticeably more indurated, commonly forming prominent sandstone ledges separated by a sharp topographic break from the transitional sandy clays which lie below in slopes continuous with the underlying Pierre Shale. In contrast, the Fox Hills beds in the type area contain more silt and clay and are relatively unconsolidated; here the little topographic separation that exists between Pierre and

Fox Hills generally occurs at or near the base of the transitional clayey silt (Trail City Member).

While the knowledge of the Cretaceous and Tertiary rocks to the west steadily increased, the pioneer work of Hayden in the area of the type Fox Hills stood unsupplemented by any new work until 1884 when Bailey Willis investigated the lignites of the Great Sioux Reservation for the U.S. Geological Survey. Willis concentrated his brief study in the western part of the Moreau-Grand divide where Fox Hills beds are exposed in the river valleys and the Hell Creek underlies the divide. On the basis of fossils he distinguished "two deposits—which have been determined by Mr. J. B. Marcou to be of the Laramie and Fox Hills formations, both of which are placed in the Cretaceous group in the present classification of the Geological Survey;" and he gave the following chart (Willis, 1885, p. 11):

Formation	Fossils	Lithology
Laramie	*Ostrea glabra,* M. & H. *Melania insculpta,* M.	Light-yellow, gray, and sandstones and arenaceous shales, with thin beds of lignite and iron nodules.
Upper Fox Hills	*Tancredia Americana,* M. & H. *Cucullaea (Idonearca) Schumardi,* M. & H.	Brownish sandstone.
Fox Hills	*Scaphites Cheyennensis,* Owen (sp.) *Turris (Surcula)? contortus,* M. & H. *Volsella attenuata,* M. & H. *Scaphites Conradi,* var. *intermedius,* M. & H.	Grayish-blue shale below brownish unfossiliferous sandstone.
Lower Fox Hills (= Fort Pierre of Hayden in part.)	*Scaphites Cheyennensis,* Owen (sp.) *Pteria(Oxytoma) Nebrascana,* E. & S. *Spironema tenuilineata,* M. & H. (sp.) *Cucullaea exigua,* M. & H.	Dark-gray to blue-black tenacious alkaline clays, locally hardened to compact clay rock.

He saw the units labelled "Upper Fox Hills" and "Lower Fox Hills" in the Grand River valley on and 15 miles east of Firesteel (then Flint) Creek, respectively; they are the *Tancredia-Ophiomorpha* biofacies of the Timber Lake Member and the fossiliferous Trail City Member in the vicinity of Bullhead. The "Fox Hills" unit is described from exposures on the Moreau near the mouth of Thunder Butte Creek in which area silty clay and shale (Irish Creek lithofacies of the Trail City) laterally replace the

Timber Lake Member but carry its marine fauna. The reference to a "brownish unfossiliferous sandstone" probably refers to the tan-weathering, silty to sandy upper part of the clay where it underlies old river terraces, for there is no sandstone bed at the position indicated in this area. Willis (p. 12) emphasizes that "the upper part of Hayden's Fort Pierre" though predominantly shaly, carries a Fox Hills fauna, and in doing so, implies that Hayden included all the predominantly shaly beds in the Pierre. Willis could have had no more basis for interpreting where Hayden put the base of the Fox Hills than we have today, but he obviously favored restriction of the name to beds of sandstone.

Willis, like Hayden, took the Colgate-type sand as the basal unit of his "Laramie formation," recognizing that in addition to its local occurrence as indurated, butte-capping, lenses, this ". . . sandstone grades over large areas into arenaceous shale or becomes soft and incoherent, and there is then much difficulty in determining its correct position." This sentence implies that he considered as lateral equivalents the Colgate lithofacies and the "arenaceous shale," which is undoubtedly the banded beds or Bullhead lithofacies; but his work was too brief to suppose that he could have had more than a general suspicion of the regional validity of this relationship. The same reference to "arenaceous shale" is his only mention of the conspicuously thin-bedded Bullhead Member, and in relating it with the Colgate sand he implies that he considered it a part of his Laramie.

Willis' work was a short study directed toward the assessment of the Laramie lignites and understandably added little more than some local detail and nomenclatural modernization to Hayden's work. The chief departure is the local application of the term Laramie to the beds of the Lignite group and the relegation of these to the Cretaceous. The informal subdivision of the Fox Hills by Willis is an interesting step that probably stems from his uncertainty in relating the Fox Hills sequence he saw on the Grand River with the markedly different aspect of the same sequence he saw farther southwest on the Moreau River.

James E. Todd, the first State Geologist of South Dakota, noted in the introduction to his study of the northwest-central part of the state that this area (the type area of the Fox Hills) is ". . . a portion almost unknown to the scientific mind" (Todd, 1910, p. 14). He then relieves this condition by presenting the most comprehensive geographic and geologic study of the area in the literature. The work, a basic one for Fox Hills stratigraphy, presents the first reasonably accurate, though crude, geologic map of the area, the first detailed stratigraphic sections, and an annotated list of fossils with locality data. The text contains many new observations clarifying the stratigraphy. It is this work rather than Hayden's that must be taken as the point of departure for modern interpretation of the type Fox Hills Formation.

Todd turned to Meek and Hayden "to ascertain the original concept of the formation" (1910, p. 28) and he followed their method of delimitation, basing the lower contact on lithology and the upper one on the change in faunas from marine to brackish. He noted that the lower part of the formation is very much like the Pierre Shale but can be distinguished by the silt and sand included in the clay and by the abundance of concretions, many of them highly fossiliferous. This general description together with his faunal lists, measured sections and descriptions of collecting localities leave no doubt that he included the Trail City beds in his lower Fox Hills and not in the Pierre Shale.

Speaking of the upper Fox Hills (1910, p. 28) he noted that it "... is not distinguishable lithologically from the Laramie, but its fossils are marine, while those of the Laramie are fresh or brackish water species." In his conscientious attempt to apply this basis of separation, Todd came very close to recognizing some of the lateral facies changes, observing that, "It seems in comparing different exposures that the transition from the marine to the fresh water form is not always in the same lithologic formation. On Grand River it seems to be mostly in sand, on the Moreau in clay." (1910, p. 29).

Todd recognized the lenticularity of individual lithologic units in the area but distinguishes a gross uniformity of sequence in the Fox Hills and Laramie beds which he presents in the form of a generalized section (Todd 1910, p. 34-35). (The descriptions are abridged; the parentheses are mine.)

H. Clays, shaly and plastic with lignite frequently developed ... the whole being very irregular in stratification, and little of it consolidated .... 300 to 350 (feet)

I. Sands with concretions and some layers of lignite ... toward the east usually form the capping of Buttes. .... Numerous fossil patches, as e.g. at Dog Butte containing *Ostrea, Corbicula, Melania, Neritina*, etc.

   With this closes the Laramie, except that in certain places Laramie fossils appear in the upper part of the next (below) stratum, which usually shows marine fossils in its lower portion. (no thickness given)

J. Sand and clay interstratified in thin layers ... with brackish water fossils in the upper layers, such as, *Corbicula, Ostrea, Melampus, Volsella*. Below appear *Natica, Dentalium, Mactra, Scaphites* .... 60 to 75

K. Yellow sandstone, much of it massive, some of it quite hard near the top, containing *Tancredia, Calista (Callista), Fusus,* etc. .... 50 to 115

L. Stratified sandy clays with several levels of concretions lenticular and globular, many very fossiliferous, though some are barren .... 100 to 160

The lower part of this last stratum shades often quite imperceptibly into the dark plastic clay of the Pierre.

Todd's subdivisions in this general section clearly mark the same four units that are recognized as the dominant lithofacies of the Fox Hills today: L = Trail City Member, K = Timber Lake Member (Todd even noted that this unit is not persistent to the west along the Grand River), J and I—respectively, the Bullhead and Colgate lithofacies of the Iron Lightning Member with the inclusion in I of some basal Hell Creek containing local sand beds of Colgate lithology. Division H is, of course, the lower part of Hell Creek. In an attempt to make his Fox Hills-Laramie contact conform with the change from marine to brackish- and fresh-water faunas, Todd is forced into stratigraphic inconsistency. Although he obviously preferred (1910, p. 34) to have the contact correspond to the base of his unit I, which is Hayden's Fox Hills-Lignite Group contact, the irregularities of faunal distribution unrecognized by Meek and Hayden compelled Todd to state (1910, p. 35) that "The junction of Laramie and Fox Hills is in J in some cases and elsewhere between J and K." Considering Todd's demonstrated stratigraphic sense it is a wonder that the indefinite nature of this contact did not provoke him into abandoning the criterion of faunal change for that of lithology.

Todd's significant study of the Fox Hills is based on field work done in 1902 in the type area and is a part of Bulletin 4 of the South Dakota Geological Survey. Although this is the report of the State Geologist for 1908 it did not appear in print until 1910. Unfortunately this was too late to be available to the U.S. Geological Survey party under Calvert that spent the field season of 1909 in the type area. The subsequent publications of the government geologists (Stanton, 1910; Calvert and others, 1914), discussed below, presented a different delimitation of the Fox Hills than that of Todd and because of their much wider circulation overshadowed Todd's work to such a degree that it has generally been ignored.

At about this same time appeared Darton's (1909) Water Supply Paper on South Dakota. The little information it offered on the Fox Hills in its type area is not based on field observations but is derived from early works of Todd. However, Darton did make a significant departure in terminology, employing the term "Fox Hills Sandstone," instead of "Fox Hills Formation," thus marking the spread of this western usage into the type area. As the chart (Fig. 4) shows, this change in terminology brought with it a corresponding change in the interpretation of the lower contact of the formation.

An Act of Congress in May, 1908, called for a survey of coal resources on the Standing Rock and Cheyenne Indian Reservations and sent a U.S. Geological Survey party into the type area of the Fox Hills in 1909 under W. R. Calvert. Calvert's (1914) report was preceded by a paper by Stanton (1910) who visited the Survey party in the summer of 1909 during a field study of the Fox Hills-Lance succession in connection with the Laramie problem. Stanton's interpretation of the "Fox Hills Sandstone" was followed by Calvert. Apparently neither author had Todd's 1910 report available to him.

Stanton concentrated his study in the western part of the type area where outcrops of the Fox Hills-Lance contact were available, but he saw much of the Fox Hills throughout the Missouri Valley area and was well aware of its considerable variation in character (Stanton 1910, p. 174). He consistently employed what might be called the Rocky Mountain definition of the Fox Hills, drawing the base of the formation at the first sandstone and including the transitional clayey silts (Trail City Member) below in the Pierre Shale. Nevertheless, he was aware that "it is by no means certain that the top of the dark shale necessarily taken as the top of the Pierre.... represents exactly the same geologic horizon. The fauna of at least the upper 100 feet of Pierre shale is essentially a Fox Hills fauna... and differs very little from that of the overlying sandstone, while it lacks all the species which in other areas are especially characteristic of the Pierre" (p. 177).

The principal contribution to local Fox Hills stratigraphy in Stanton's 1910 paper is his clarification of the upper contact of the formation. By careful collecting of fossils Stanton revealed that the brackish-water fauna in the light-gray sandstone (Colgate lithofacies), which previously had been taken as the base of Lignite Group by Hayden (i.e. Hayden's "Bed Q") and included in the Laramie by both Willis and Todd, contained marine elements—including ammonites—of the underlying Fox Hills fauna. On this basis he included the sandstones with the mixed brackish and marine fossils in the Fox Hills, placing the Fox Hills-Lance contact at the top of the bed in which the highest marine fossils are found. It does not seem to have occurred to Stanton that he was using gross lithology to define his Pierre-Fox Hills contact and

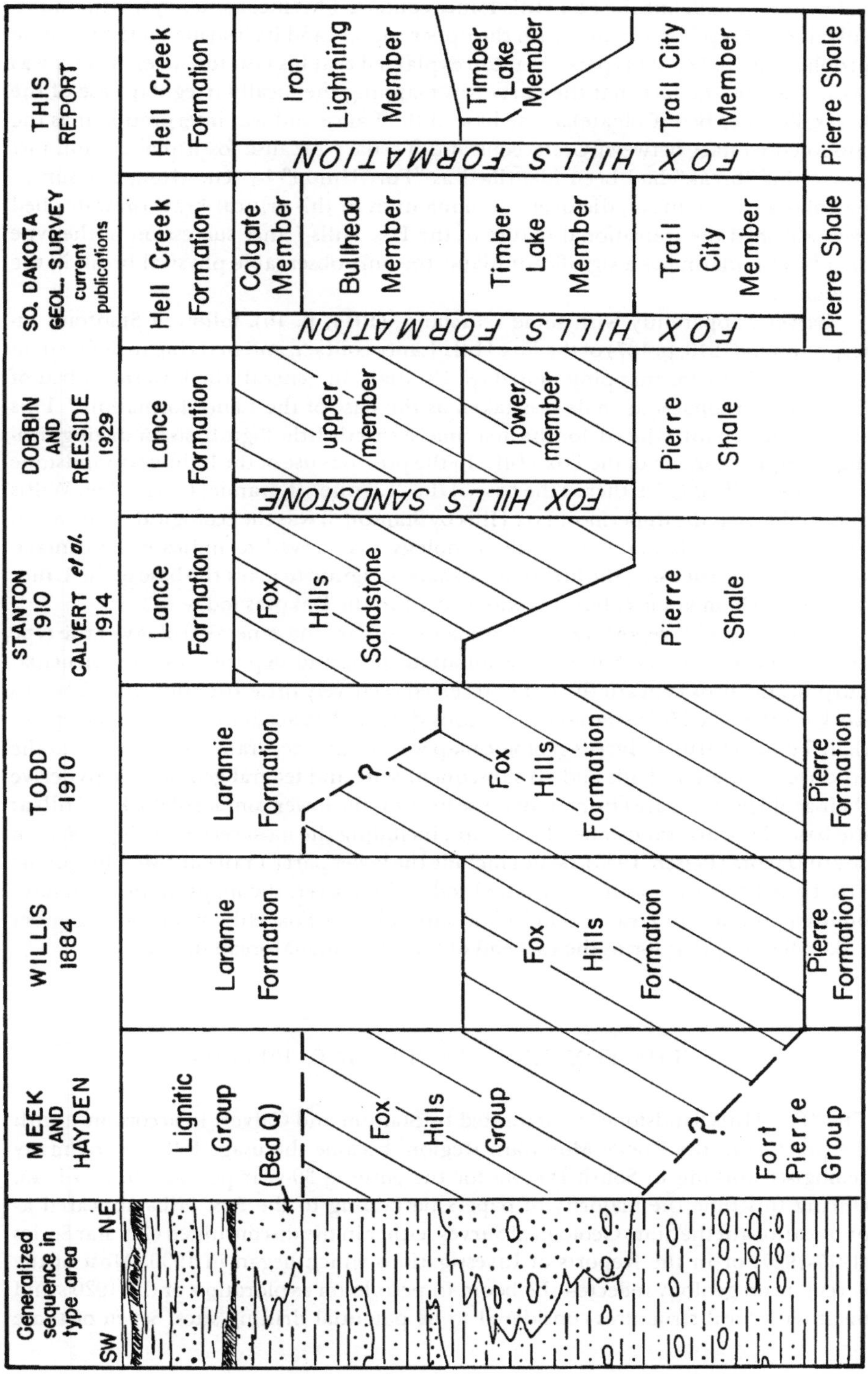

FIG. 4. Development of classification and nomenclature of the Fox Hills Formation in the type area.

paleontology to define the Fox Hills-Lance contact. This inconsistency was undoubtedly due to his preoccupation with the upper contact and its bearing on the Laramie
problem. On p. 181 of his paper Stanton explained that his visit to Calvert's party was
occasioned by the fact that the party was mapping the locally irregular base of the
brackish-water bed (Colgate) as the base of the Lance and was interpreting it as the
supposed contact between Cretaceous and Tertiary because fossil plants from just
above this "break" had been identified as "Fort Union" by Knowlton, the Survey
paleobotanist. Stanton's discovery of ammonites in this critical brackish-water bed
re-established the transitional nature of the Fox Hills-Lance succession in the type
area by eliminating as a significant hiatus the only observable physical break in the
sequence.

Calvert, apparently with some reluctance (1914, p. 16), followed Stanton's interpretation (1910, p. 187) of the Fox Hills-Lance contact, and in trying to define it on
a practical basis for mapping, noted (p. 18) that "In general, the lowermost bed of
lignite or carbonaceous shale was taken as the base of the Lance formation." This
significant step introduced, for the first time, a strictly lithologic basis for distinguishing the upper contact of the Fox Hills. In the previous use of the light-gray sandstone
of Hayden's "Bed Q," either as the base of the "Lignite" (Laramie) by Hayden, Willis
and Todd, or as the uppermost Fox Hills by Stanton, it was the ecological implication
of the contained fauna and not the lithology that served to indicate the contact.
Calvert's device of using the first lignitic shale or lignite to mark the base of the Lance
did not catch on at once, but it is widely used in the Dakotas today.

Calvert's (1914) report was the last major work on the general geology of the type
area of the Fox Hills. Since the evaluation of lignite deposits was his principal
purpose, the beds beneath the Lance received relatively little attention and Calvert's
work on the Fox Hills stratigraphy is considerably less accurate and less perceptive
than Todd's (1910). Calvert's geologic map was the first accurate surveying job in the
type area of the Fox Hills and all subsequent state and federal geologic maps derive
their geology of this area from it. Because he drew the lower contact of the Fox Hills at
the base of the first sandstone, these maps, including the most recent geologic map of
South Dakota (Petsch, 1953), do not include the lower part (Trail City Member) of the
Fox Hills Formation as it is delimited today. Moreover, the maps in question have
omitted an outcrop area of about 40 square miles of Fox Hills sandstone (Timber
Lake Member) that forms the east end of the Cheyenne-Moreau divide.

THE FOX HILLS SANDSTONE, 1915-1945

The "Fox Hills Sandstone" as delimited by Stanton and Calvert in accordance with
the practice in the Rocky Mountain Region, became the usage followed by many
geologists working in South Dakota for the ensuing 30-year period, 1915 to 1945.
During this time the majority of papers pertaining to the Fox Hills appeared as
publications of the State Geological Survey, either in the discontinued Circular Series
(1917-1927) or in the Reports of Investigations which began in 1930. Most of the
Circulars on geology reflected the boom in petroleum exploration of the 1920's and
dealt with local structures considered to be potential drilling sites. Seven of those

Circulars described small parts of the type area of the Fox Hills (Wilson, 1922, 1925; Wilson and Ward, 1923; Ward and Wilson, 1922; Russell, 1925a, 1925b, 1926) and contributed to local stratigraphic details. Most of these were in the western part of the type area and involved the upper Fox Hills and overlying Hell Creek, with which they dealt in variable fashion: some avoiding any specific classification, some using the term Fox Hills Sandstone of Stanton and Calvert, and some the term Fox Hills Formation. None contributed significantly to the classification or nomenclature of the formation.

Among the Reports of Investigations those by Searight on the coal resources dealt with the Fox Hills. One of these (Searight, 1931) treated a portion of the Moreau-Grand divide where only the upper part of the Fox Hills succession is exposed. In describing these upper beds Searight (p. 4) divided them informally into a lower "banded member" and an "upper member composed of massive sandstone."

Publications other than those of the State Survey dealt only indirectly with the type Fox Hills. Thom and Dobbin (1924, p. 490), in a survey of the Cretaceous-Tertiary sequence in Montana and the Dakotas, set the stage for local use of the term Colgate with the following statement: "That the fluviatile basal sandstone of the Lance of central Montana is stratigraphically equivalent to the Colgate sandstone and to the upper white Sandstone of the type Fox Hills the writers feel confident." But use of the name Colgate for this upper sandstone—Hayden's "Bed Q" and Stanton's brackish-water bed—did not gain currency in South Dakota until the 1950's. Thom and Dobbin (1924) followed the Stanton-Calvert usage of Fox Hills Sandstone, as did Dobbin and Reeside (1929) in a paper that dealt briefly with a locality in the type area as part of a regional study of the Fox Hills-Lance contact.

Thus, in spite of increased activity by the State Geological Survey and continued interest in the "Laramie problem," the 30 years between 1915 and 1945 saw no significant changes in the delimitation or nomenclature of the type Fox Hills sequence beyond the work of Stanton and Calvert. The informal subdivision of these beds into a lower massive, yellowish-brown sandstone, a middle "banded" unit of thin-bedded shale and sandstone, and an upper gray-white sandstone unit, was commonly, if not unanimously, accepted and the whole was called the Fox Hills Sandstone. In the continuation of the outcrop from the type area into the Missouri Valley of North Dakota the same threefold subdivision was utilized, but here the name Colgate Member was formally used for the upper sandstone as early as 1942 (Laird and Mitchell, 1942).

THE FOX HILLS FORMATION, 1945 TO THE PRESENT

The first step toward the current classification of the type Fox Hills was taken by Morgan and Petsch (1945) when they removed the transitional unit of fossiliferous clayey silt underlying the lower massive sandstone of the Fox Hills from the Pierre shale and reinstated it as the basal member of the Fox Hills. This return to the earlier practice of Todd was enforced by giving the formal name Trail City Member to this basal unit, and the name Timber Lake Member to the overlying sand. In the same work Morgan and Petsch (1945, p. 12) returned to the usage Fox Hills Formation, discarding the name Fox Hills Sandstone on the grounds that it is misleading as applied to a sequence of such variable lithology.

The work of Morgan and Petsch was confined to the south side and eastern end of the Moreau-Grand divide in an area where the upper part of the Fox Hills was absent or only partially present. It remained for other workers to complete the formalization of the upper divisions of the formation. This was done in the State Geological Survey series of Quadrangle Maps on the geology of the coal lands, begun early in the 1950's. Curtiss (1952) was the first to utilize the name Colgate Member for the upper gray-white sandstone in the type area, and Stevenson (1956) introduced the name Bullhead Member for the thinly-interbedded shale and sand unit that Searight (1931) had informally called the "banded member."

To date, ten maps of the Geologic Quadrangle series published by the State lie within the type area of the Fox Hills; seven of these are accompanied by a description of the geology (Curtiss, 1952, 1954; Pettyjohn, 1961; Stevenson, 1956, 1959, 1960a and 1960b). The fourfold subdivision of the Fox Hills (Trail City—Timber Lake—Bullhead—Colgate) has been mapped in the latter five of these quadrangles, all of which lie in eastern Corson and central eastern Dewey counties. With two exceptions the members have not been used as Fox Hills subdivisions outside the type area of the formation. Colgate is a term applied rather loosely throughout eastern Montana, eastern Wyoming and the Dakotas to any gray-white sandstone near the Fox Hills-Hell Creek (Lance) contact. The other exception is the local use of the names Trail City and Timber Lake in the Missouri Valley region of North Dakota by Fisher (1952).

The U.S. Geological Survey, in two resource reports pertaining to parts of the type area (Denson, 1950; Tychsen and Vorhis, 1955), has followed the Stanton-Calvert usage of Fox Hills Sandstone—with the exclusion of the Trail City beds that this classification implies. Although neither of these reports was primarily concerned with the Fox Hills beds, they perpetuated the "Fox Hills Sandstone" classification and its threefold subdivision. In the Cretaceous correlation chart (Cobban and Reeside, 1952, chart 10b, column 92) the term Fox Hills Sandstone is used to embrace the four-member subdivision of the State Survey, thereby introducing a third style of nomenclature for the formation in its type area.

4. REVISION OF SUBDIVISION

The concept of the type Fox Hills as a formation made up of four successive members is an oversimplification of fact. The four named members in use at the present time do indeed apply to four lithogenetically distinctive parts of the formation but these are neither uniformly successive nor are all of them continuous within the type area. Two of the four members (Bullhead and Colgate) are so intricately interrelated that their separation obscures rather than clarifies their true relationship. All of the lithogenetically different bodies of sediment that make up the type Fox Hills are important to this study—concerned as it is with environments—but not all of them useful as map units deserving formal stratigraphic names. Only minor modification of the existing terminology of the type Fox Hills is necessary to represent the stratigraphic details with more precision and to provide more widely applicable subunits for geologic mapping.

The relationships of the principal Fox Hills lithofacies are illustrated schematically in Fig. 5. This diagram together with the brief stratigraphic summary, pages 14 to 15, suffice to show in a general way the relation of the revised nomenclature to the

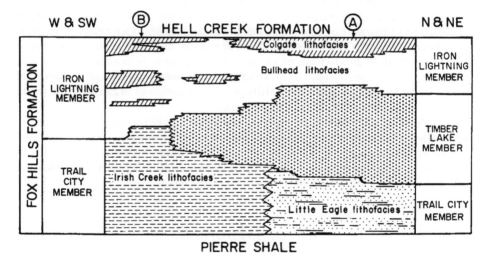

FIG. 5. Revised subdivision and nomenclature applied to the principal lithogenetic units in the type Fox Hills Formation. The lithofacies terminology is informal.

stratigraphic facts; the details follow in the section on stratigraphy. It is evident from Fig. 5 that the terminology of Fox Hills subdivisions in a section taken at A, where Trail City, Timber Lake, Bullhead and Colgate subdivisions follow one another in unbroken sequence, is of little use in describing the formation as it occurs at locality B. The complicating factors are 1) the southwestward termination of the Timber Lake

45

sand body, 2) the westward change in the nature of the Trail City, and 3) the westward thickening of the upper part of the formation with the concomitant appearance of Colgate lithology at more than one horizon.

The most obvious and internally consistent features of the formation are the persistent lower part of clayey silt, the westward-tapering sand wedge, and the varied upper beds which overlie both the sand wedge and, beyond its termination, the clayey silts. The Trail City Member is retained and redefined to include all of the lower clayey silt unit, not just the highly fossiliferous part. The Timber Lake Member is retained for the sand body, much as originally defined. The closely related Bullhead and Colgate units are combined as a single member—the Iron Lightning Member. This subdivision provides three mappable units which can be used as Fox Hills subdivisions wherever it is desirable to map in stratigraphic detail in the type area and much of the Missouri Valley outcrop area. Only in the eastern part of the type area do all three members occur together as individually mappable units. To the west the Trail City and Iron Lightning Members comprise the entire Fox Hills. In North Dakota only the Iron Lightning and Timber Lake are potentially useful subdivisions as the latter drops in section at the expense of the Trail City.

A broader, natural subdivision of the Fox Hills into lower and upper parts is afforded by its major environments of deposition. The Trail City and Timber Lake Members form a lower part of off-shore marine deposits and the Iron Lightning Member an upper part of near-shore marine and brackish-water deposits. In areas peripheral to the type Fox Hills, where one or more of the members begins to lose its identity, a twofold subdivision of the formation apparently remains obvious and might be useful in mapping. In describing the type area, adoption of the informal subdivisions lower and upper Fox Hills provides a means of organizing the stratigraphic data in an environmental framework.

In addition to the formal members of the Fox Hills Formation, informal units, here called lithofacies, are used to single out lithogenetic units essential for environmental reconstruction. As used in this report, a lithofacies is one or more bodies of sediment or sedimentary rock distinguished from enclosing deposits by noteworthy lithologic, organic and/or internal structural characters.

USE OF LITHOFACIES

Wells (1947, p. 119) introduced the term lithotope for what is here called a lithofacies—the rock record of a particular environment. As both Dunbar and Rodgers (1957, p. 137) and Teichert (1958, p. 2730) have made clear, the term lithotope has subsequently been applied to areas of uniform environment rather than to rock bodies. This unfortunate redefinition has prevailed in usage to the point where the original sense of the term seems unredeemable. Among other informal terms in use, lithofacies is apparently the only one flexible enough, because of persistent loose usage, to encompass the definition given above.

The definition of lithofacies employed here is one of several currently in use. Among these existing definitions there is a wide range of flexibility and utility; some are largely conceptual and difficult to use in practice, others have their usefulness impaired by the addition of unnatural restrictions as corollaries. The definition of lithofacies above borrows some of its phraseology from a similar but longer one by Moore (1957, p. 1783), who stated that "lithofacies might be defined as any part of a stratigraphic unit (or arbitrarily delimited body of sedimentary deposits) that is distinguished by noteworthy lithologic characters differing from those of other parts of the unit (or arbitrarily delimited body) . . ." Here, the idea of a lithofacies as a discrete rock body seems implicit, but in the ensuing discussion Moore (p. 1783-1784) introduced the restriction of laterality—the lithofacies can only be a lateral variant of contemporaneously formed strata. In the same article Moore (p. 1787) accepted the term lithosome as informal nomenclature for a body of sediments distinguished by lithology. The distinction between lithosome and lithofacies, while not made clear, is apparently laterality, a requirement of lithofacies but not of lithosome. Nevertheless it is hard to see any practical difference between Moore's use of lithofacies to "characterize particular sedimentary environments" (p. 1785) and his use of lithosome as the "stratigraphic record of a more or less uniform lithotope . . ." (p. 1788).

Wheeler and Mallory (1956) presented a clear distinction between lithofacies and lithosome. Briefly, they restricted lithofacies to ". . . statistically derived lateral lithic variants of a vertically segregated stratigraphic interval" (p. 2722), following the usage of Sloss et al. (1949), and gave lithosomes considerable scope as "verticolaterally segregated units" (p. 2719). Except for limiting the stratigraphic relationships of lithosomes to intertonguing, their use of this term closely approximates the definition of lithofacies used in the present report. In fact, Wheeler and Mallory (1956, p. 2720) stated that were it not for the restriction of the term lithofacies to the definition by Sloss et al. (1949), "there would have been no need for the term lithosome."

The contention here is that there is no valid reason to restrict the term lithofacies solely to the usage of Sloss et al. (1949). The map in Fig. 22 showing the distribution of Fox Hills lithofacies at the time of accumulation of the *Cucullaea* Assemblage Zone does not differ significantly in kind from a lithofacies map of the Upper Cretaceous rocks of the western interior. In the former, where the resolution is down to an interval of sediment a few feet thick that approximates a single time plane, lithofacies limits can be pinpointed and there is no need to define them arbitrarily. The greater the thickness of beds and of the time interval to be represented in the construction of a lithofacies map the greater the generalization needed to depict the shifting of lateral boundaries as a single line. Sloss and others have used a simple statistical device as a method to accomplish this generalization. These extremes of usage may seem too obvious to warrant spelling out, yet in the literature on stratigraphic terminology the statistical device, which is simply methodology, too often is erroneously taken as part of the concept. Similarly, the strict laterality imposed by an investigator in choosing a time-bounded sequence of rock as a practical (operational) means for generalizing areal lithofacies distribution should not be mistaken for a characteristic of the several lithologically distinct units he is summarizing in his two-dimensional lithofacies map. The means an investigator chooses to select, employ or depict lithofacies to best meet his needs should not be confused with the definition of the natural, three-dimensional feature itself.

In its broad sense then, a lithofacies has no restrictions either of laterality or of size. It may consist of a single rock body or more than one rock body, like the Colgate lithofacies of this study. The formal rock-stratigraphic units, at least those distinguished by their lithologic unity, are, of course, lithofacies by definition. In certain contexts it is more meaningful to refer to a formal unit, for example, the Timber Lake Member of the Fox Hills, as the Timber Lake lithofacies; one situation in which such interchangeability is useful is where the lithologic identity of a named unit extends beyond its practical mappable limit. The large, arbitrarily delimited lithofacies with statistically derived lateral boundaries and temporally defined vertical cut-offs are a somewhat special case, their very size requiring generalization of natural limits. In light of the increasing interest in environmental and ecologic reconstruction it is predictable that the term lithofacies will find as great use in detailed stratigraphic studies as it has in broad regional syntheses. Its convenience in detailed stratigraphy merits its preservation in a simple, flexible and objective sense. Such usage does not compete with formal stratigraphic nomenclature and, as Weller (1960, p. 521) points out, it can be ignored by the investigator not concerned with such details.

Geographic names are used to designate the lithofacies employed in this report. This practice is adopted chiefly for convenience of reference. Place names avoid implications as to environmental interpretation and their use fits in with the "type section" methodology which is as useful for reference to lithofacies as it is for reference to formal lithostratigraphic units. Lithologic names available for lithofacies are too limited in number to permit enough distinctions without resorting to unwieldy descriptive phrases. Use of the same method as that applied to formal nomenclature makes possible uniform reference where a particular lithofacies also has a formal name (Timber Lake Member—Timber Lake lithofacies). In the present report, the expedient of retaining the old member names Colgate and Bullhead for the lithofacies to which they were originally applied illustrates the flexibility of the method.

5. DESCRIPTIVE STRATIGRAPHY

THE UPPER PIERRE SHALE

The Pierre Shale is intermittently exposed beneath the Fox Hills outcrop all along the south side of the Cheyenne-Moreau divide, along the Moreau River valley as far west as Irish Creek, and along the Grand River valley as far west as Bullhead. Away from the rivers the outcrop is usually covered with grass and in the river breaks it is commonly slumped; rarely is it possible to piece together a reliable stratigraphic section of any appreciable thickness. Nevertheless, the three upper members of the Pierre Shale—the Virgin Creek, Mobridge and Elk Butte—all have their type localities in or closely adjacent to the type area of the Fox Hills. Only the Mobridge and Elk Butte members are pertinent to the study of the type Fox Hills; the Mobridge because it contains the first macrofossils beneath the Fox Hills, including the important regional Range Zone of *Baculites clinolobatus,* and the Elk Butte because it is gradational into the Fox Hills and not simple everywhere to separate from it.

In the original subdivision of the Pierre Shale (Searight, 1937) the upper three members were distinguished primarily on the basis of the calcareous nature of the Mobridge Member and the non-calcareous nature of the Virgin Creek and Elk Butte shales, respectively below and above it. Although most of Searight's subdivisions of the Pierre Shale have undergone revision (Gries, 1942; Crandell, 1950) the upper three members survive as originally defined. This is probably due to the fact that the Mobridge Member is a rather distinctive, calcareous to chalky, light-gray shale in the southern half of the Missouri Valley of South Dakota, where detailed studies of the Pierre Shale happen to have been concentrated. In a recent study of the stratigraphy and micropaleontology of the upper three members of the Pierre Shale in the Moreau and Grand River valleys, Mello (1962, unpublished) found that these members could be distinguished only in a very gross way because of the reduction and local stratigraphic fluctuation of the calcareous content of the Mobridge. The generally poor exposures and an apparent lack of persistent key beds render the Elk Butte Member particularly difficult to handle stratigraphically in any detail, but Mello was able to establish the persistence of two bentonites in the upper part of the Mobridge which are most helpful in working out the stratigraphic and faunal relationships of the Upper Pierre and Fox Hills (Fig. 6).

Pertinent features of the upper Pierre Shale are shown diagramatically in Fig. 6. The two key bentonite beds, between 50 and 60 feet apart, are in the dominantly calcareous upper 100 feet of the Mobridge Member. Within an interval of about 20 to 50 feet above the upper key bentonite the calcareous shale gives way to the non-calcareous shale of the Elk Butte Member. Commonly this broad zone of transition contains small, rusty-weathering calcareous concretions that break up into shards on the outcrop. Throughout the area the upper part of the Mobridge Member is a gray to dark-gray color similar to that of the overlying Elk Butte Member and lacks the light-gray to buff weathered surface that serves to distinguish it from adjacent units farther south in the Missouri Valley area.

49

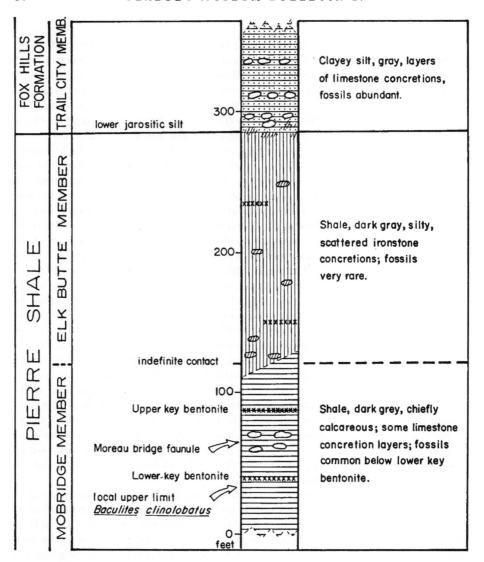

FIG. 6. Diagrammatic section of the upper part of the Pierre Shale in the type area of the Fox Hills Formation.

Macrofossils in the Mobridge Member are confined chiefly to the beds below the lower key bentonite. Here they are not uncommon, although rarely abundant. Concretionary fillings of the body chambers of baculites are locally a plentiful fossil in the lower part of the Mobridge. Their rather abrupt disappearance a short distance beneath the lower key bentonite appears to be a widespread feature and marks the end of the indigenous baculites in the region. *Baculites clinolobatus,* the principal species in this baculite assemblage, marks a widespread range zone in the Upper Cretaceous of the western interior (Cobban, 1958, p. 114); it is a very useful zone in determining the stratigraphic position of the Fox Hills Formation relative to the Pierre Shale through much of the northeastern Great Plains region. The *B. clinolobatus* Range Zone terminates approximately 250 feet below the base of the type Fox Hills.

Above the lower key bentonite macrofossils are rare but in the bluffs of the Moreau River in the vicinity of the South Dakota Highway 63 bridge, a sparse fauna of scaphitid ammonoids and pelecypods, referred to here as the Moreau Bridge faunule, was found in small limestone concretions between the lower and upper key bentonites.

The Elk Butte Member is poorly exposed everywhere and the apparent lack of distinctive key beds makes difficult the piecing together of measurements from small outcrops to form trustworthy composite sections. This is certainly true at the type locality (Searight, 1937, p. 50) where partial, disconnected exposures on the south flank of the upland capped by Rattlesnake Butte (mistakenly called Elk Butte by Searight) in T. 20 N., R. 29 E., do not permit reconstruction of a reliable type section. At several localities along the Moreau River a thin bentonite is present about 50 to 55 feet below the Fox Hills contact and in the same area Mello notes (1962, unpublished) a second thin bentonite in the lower part of the member between 60 to 70 feet above the upper key bentonite of the Mobridge. How persistent these are throughout the area as a whole is not known. Calculations of the total thickness of the Elk Butte Member range from 160 to 210 feet and indicate a gradual thinning toward the southwest.

The characteristic lithology of the Elk Butte is a gray-weathering, non-calcareous, very finely silty shale that breaks down into tiny chips on weathering. The fresh shale when excavated is dominantly blocky and fracture surfaces are commonly coated with brownish to purplish-gray stain. Reddish-brown to purple-weathering sideritic concretions occur in the shale, and at places reddish-brown-weathering limestone concretions are found. The silt content of the shale is somewhat variable but there is a gradual increase in this constituent in the upper 50 feet of the unit. Here the silt appears as scattered zones of laminae and small pods in the shale, and nearer the Fox Hills contact the shale itself becomes more silty.

Macrofossils are very rare in the Elk Butte Member, with the exception of small linguloid brachiopods that are locally fairly plentiful. Uncommonly, single ammonoids were found in some of the calcareous concretions in the upper part of the member but the scant number of these, two fragments in four field seasons, merely accentuates the general absence of a normal marine fauna in the Elk Butte.

The local Moreau Bridge faunule assumes some importance in the area of the type Fox Hills as the only macrofossil accumulation known in the Upper Pierre Shale above the lower bentonite of the Mobridge Member. It occurs sparsely scattered in small, hard limestone and/or gypsum-coated, punky, concretions between 15 and 30 feet beneath the upper key bentonite. Scaphitids of several kinds and the so-called *Inoceramus*, *I. fibrosus* (Meek and Hayden), are the more common elements but it also includes a true *Inoceramus* whose species is not determinable from specimens found thus far, two genera of relatively large woodboring bivalves (probably *Opertochasma* and *Turnis*), and rare protobranch bivalves. The scaphitids include early variants of *Scaphites* (*Hoploscaphites*) *nicolleti* (Morton), and at least two other undescribed species that also appear to be ancestral to Fox Hills stocks.

LOWER FOX HILLS FORMATION

MAJOR FEATURES

The Lower Fox Hills, like the formation of which it is an informal subdivision, is a heterogeneous stratigraphic unit. It consists of silty and sandy clay, clayey silt and sand, silt, fine-grained sand and some silty shale. Limestone in the form of ovoid concretions and tabular concretionary masses is a minor but conspicuous and characteristic constituent. Some ferruginous cementation and abundant rusty stain on weathered surfaces give the sandy parts of the unit their diagnostic light yellowish-brown color. The silty and clayey parts of the Lower Fox Hills weather light gray or light yellowish-gray except in the fresher cuts along the rivers where the color may be as gray as the upper part of the Pierre Shale, a superficial resemblance which has led a number of workers to include these clayey beds in the Pierre Shale.

The relationship of the lithofacies that make up the Lower Fox Hills is shown in Fig. 5. In broad terms the Lower Fox Hills in the type area can be regarded as a unit of clayey silt, the Trail City Member, with a northeastward-thickening sand wedge, the Timber Lake Member, in its upper part. The Trail City clayey silt is itself differentiable into eastern and western lithofacies on the basis of marked changes in bedding, fossil content and other secondary characteristics. These are designated, respectively, Little Eagle and Irish Creek lithofacies. The highly fossiliferous concretion layers that are the most conspicuous characteristic of the Little Eagle lithofacies all lose their fossil assemblages within a relatively short distance laterally of one another so that it is possible to set an easily recognizable, if arbitrary, western limit for it.

Both the Little Eagle lithofacies and the lower part of the Timber Lake Member grade southwestward into the Irish Creek lithofacies of the Trail City. The upper part of the Timber Lake is in abrupt but conformable contact with the Upper Fox Hills, the base of which descends stratigraphically southwestward. Although the Timber Lake Member thickens eastward and northeastward at the expense of both the Irish Creek lithofacies and the Upper Fox Hills, the Little Eagle lithofacies maintains a fairly constant thickness within the type area. However, the Trail City as a whole becomes sandier eastward, and northeastward beyond the type area it grades into the Timber Lake lithofacies to such an extent that only a few feet of it are present on the east side of the river in the southern Missouri Valley region of North Dakota.

The Lower Fox Hills ranges from 175 to 220 feet in thickness. The 6 or 7 composite sections that permit an estimate of total thickness are too few to reveal any pronounced trends. In general, thicknesses of 200 feet or more are prevalent in the central and western part of the area whereas thicknesses in the northeastern part are under 200 feet. Apparently the northeastward thickening of the Timber Lake Member is more than compensated for by northeastward thinning and southwestward thickening of the Trail City Member.

KEY BEDS

The Lower Fox Hills contains a sequence of about 20 key beds that do not vary in stratigraphic position relative to one another but do vary considerably in their geographic distribution and lateral continuity. Consequently all key beds in the succession do not occur at any one locality; a number of them are restricted largely to one lithofacies and others to limited areas within one or more lithofacies. Many can be traced from one lithofacies to another and these have supplied the primary control for establishing the relationships of the lithofacies. The chart (Figure 7) shows the order of succession of the key beds. Not every potential key bed is shown; some are of limited usefulness either because they are very local in distribution or because they do not permit as detailed a subdivision of the stratigraphic column as others. Figure 7 includes those key beds that experience indicates to be the most useful in working out the stratigraphy. The key beds differ so markedly in kind that separate discussion of each principal type is warranted.

BENTONITES

Thin beds of soapy, yellow-green to green-gray bentonite are not uncommon in the Lower Fox Hills, but they are generally restricted in distribution to the Irish Creek lithofacies. All the beds are thin, ranging from a fraction of an inch to as much as 8 or 10 inches in thickness. Most show some mixture with the silty clay or clayey silt matrix. Punky, white-weathering, cone-in-cone concretions are usually associated with the bentonites and are a valuable guide to their position where the bentonite is very thin or where it has graded laterally into bentonitic shale.

Four separate bentonite beds, designated A, B, C, and D in ascending order, are known in the Irish Creek lithofacies where they are relatively persistent. The position of these beds relative to the Little Eagle lithofacies and Timber Lake Member was worked out along the Moreau River in the area between State Highways 65 and 63, where the lateral transition from the Irish Creek lithofacies to this succession takes place. Here the bentonites are intermittently present, but east of Highway 63 they are rarely found. Disappearance of the two upper bentonite beds into the Timber Lake sand can be explained by dissipation of the ash falls by currents in the more turbulent environment indicated by this lithofacies. Disappearance of the two lower bentonite beds into the Little Eagle lithofacies of Trail City Member can be attributed to the reworking of the clayey silts by burrowing organisms that gives this lithofacies its characteristic mixed structure. Locally within the Irish Creek lithofacies bentonite beds were observed riddled with burrows and partially dissipated into the surrounding matrix.

In the Grand River valley bentonites are not as common in the Irish Creek lithofacies as they are along the Moreau. This could be due to the fewer good exposures of the lithofacies along the Grand but it may also be an original feature resulting from greater reworking of sediment by currents or, less likely, by organisms in this area.

JAROSITIC SILT LAYERS

Layers and pods of powdery to crusty yellow jarosite are common in the clayey silt portions of the Lower Fox Hills. In addition to the zone at the base of the Fox Hills there are two other persistent zones of jarosite concentration. One of these, here called the medial jarositic silt, is a highly silty to sandy interval locally as much as 12 feet

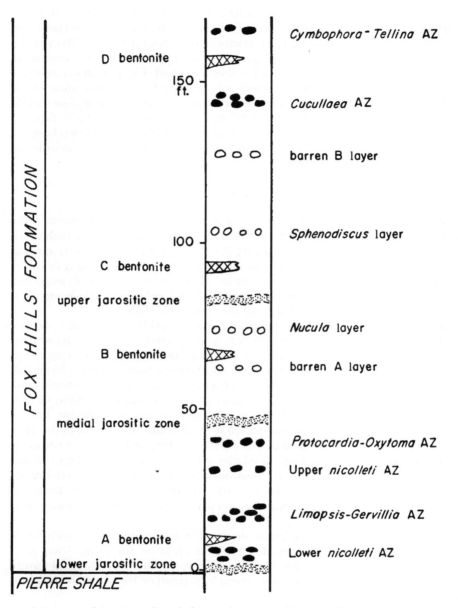

FIG. 7. Sequence of the principal key beds in the lower part of the Fox Hills Formation. (AZ= Assemblage Zone)

thick that overlies the *Protocardia-Oxytoma* Assemblage Zone. The jarosite is generally scattered throughout the sandy silt in pods and streaks but locally it is concentrated in a thin layer 2 to 6 inches thick.

The third persistent jarosite accumulation occurs conveniently near the change from the Trail City clayey silt to the Timber Lake sand in the eastern part of the type area and is also a conspicuous marker within the Irish Creek lithofacies to the west. It consists of one to three layers of highly jarositic silt or sand each 2 to 3 inches thick. At some places these thin layers are separated from one another by several feet of silty to sandy clay, but more commonly they are bunched together only a few inches apart.

The persistence of the jarositic layers—in particular the upper one, which is the more persistent of the three—is noteworthy. Because other accumulations of jarosite in addition to the key beds occur locally in the Trail City, and because no jarositic silt has characteristics peculiar to itself alone, the jarosite key beds must be identified by their relationship to other key beds. It so happens that each key jarositic silt is in a part of the Lower Fox Hills that contains a highly distinctive sequence of key beds; their position at the base, middle and top of the Trail City Member in the eastern part of the type area make them particularly useful in mapping and lateral tracing across the lithofacies change.

Samples from each of the three jarositic zones were submitted to Dr. Ernst Bolter, formerly research geochemist in the Department of Geology, Yale University, for X-ray analysis. Dr. Bolter (written communication, 1963) confirmed that the material was indeed jarosite and was associated with montmorillonite and quartz in each of the samples.

A close stratigraphic association between jarosite concentrations and glauconite concentrations indicates a relationship of some kind between these two minerals that deserves further geochemical study. The term glauconite is used here in its broadest sense because Dr. Bolter, in his analysis of the sample from the upper jarositic layer, noted a dark green mineral whose optical properties agreed with those of glauconite but which, when magnetically separated from the other minerals and subjected to X-ray analysis, gave none of the peaks reported for glauconite. The medial and upper jarositic silt zones can be traced laterally eastward in both the Moreau and Grand River areas into beds of glauconitic silty sand. The medial jarositic silt zone above the *Protocardia-Oxytoma* Assemblage Zone grades eastward and northeastward into a fine-grained sand with abundant glauconite or into a sandy silt with one or more thin beds of greensand. Thin beds of greensand continuous with the upper jarositic silt are locally present instead of the jarosite. In general the jarosite appears to take the place of glauconite where the clay-silt content exceeds the fine-grained sand content, and the reverse, but this is only a field observation and needs substantiation. Only the basal jarosite zone appears to have no glauconitic lateral counterpart within the type area.

Although not indicated in the informal name of the jarositic silt zones, the glauconitic extensions are included as part of these key beds. Where present they commonly are conspicuous on the outcrop because they weather to a rich rusty-brown or reddish-brown color.

CONCRETION LAYERS

Calcitic limestone concretions are a common feature of the Lower Fox Hills. Many are distributed in continuous layers, some of which appear to persist throughout the type area. From layer to layer, characteristics of the concretions vary somewhat so that it is possible to recognize a sequence of different types that has some lateral continuity. However, concretion type may also vary laterally within a layer, particularly where it passes from one lithofacies into another. Lithologic and structural characteristics of concretions are somewhat difficult to use as a stratigraphic tool because, with one or two exceptions, the differences involved are slight. Far more useful in the identification of many individual concretion layers are the characteristic fossil assemblages they contain. These assemblages are described in the following section.

Although none of the concretion layers appear to be absolutely barren, several are commonly unfossiliferous, yet because of their persistence are useful members of the sequence of key beds. The key beds indicated in Fig. 7 as the barren A and *Nucula* concretion layers are examples. Neither of these beds is strictly barren, in fact the lower one locally carries a very sparse but distinctive faunule consisting chiefly of nuculid bivalves. On most outcrops, however, fifty or more concretions from either of the two layers could be broken open without finding a fossil.

For the local joining of separate measured sections concretion layers are the most useful marker beds. They are common and can usually be traced across grassed areas between exposures. The extent of individual layers varies greatly; some are limited to areas of a square mile or less, while others cover hundreds of square miles.

Scattered concretions or widely spaced concretions in poorly defined layers are also common in the Lower Fox Hills. Scattering is more common in the Irish Creek lithofacies, where relatively few concretion layers are discernible. The Little Eagle lithofacies of the Trail City Member has the most pronounced and abundant layering of concretions and even many of its scattered concretions are confined to well-defined zones. Concretion distribution in the Timber Lake Member is more erratic, and laterally continuous layers are not as common as in the Trail City.

ASSEMBLAGE ZONES

Todd (1910, p. 31) appears to have been the first investigator in the type area of the Fox Hills to be impressed by the fact that the successive fossiliferous concretion layers contain distinctive assemblages of macrofossils. This feature is most obvious in the concretions of the Little Eagle lithofacies of the Trail City Member, but it is also found to a less spectacular degree in the Timber Lake Member. Commonly a particular species of pelecypod dominates the assemblage, constituting as much as 95 percent of the total number of specimens in a concretion. Other concretion zones are characterized by more variety, but the dominance of one or two species is marked. In a few zones, assemblages locally consist principally of ammonoids or, less commonly, gastropods. The bivalve fauna of the Lower Fox Hills as a whole is relatively uniform and the great majority of the species range throughout the unit; the marked differences in the faunas of successive zones result from differences in the dominant species.

Zones of this type fall within the category of Assemblage Zones (Am. Comm. Strat. Nomenclature, 1961, Art. 21) and are named for one or more of the dominant elements in their fossil assemblage. As Article 21 of the Code stresses, the assemblage zone has nothing to do with the range of the fossil(s) for which it is named. As will be discussed more fully in a later section, the assemblage zones of the Lower Fox Hills owe their origin to ecologic factors—probably periodic mass killings with little subsequent transport of shells. Although they are biostratigraphic units in a broad sense, and fulfill the requirements for an assemblage zone in all respects, they also qualify as key beds under the classification of rock-stratigraphic units (A.C.S.N., 1961, Art. 8, Art. 19e). They are employed here in both contexts. As key beds they are useful in working out the stratigraphic relationships within the Fox Hills Formation. As biostratigraphic units the fossil assemblages reveal ecological conditions and are useful in environmental reconstruction.

The assemblage zones of the Lower Fox Hills may consist of a single concretion layer, several adjacent layers, or an interval containing scattered concretions. Similarly a single assemblage zone may consist of a single concretion layer in one part of the area and more than one layer in another. Even in the most abundantly fossiliferous assemblage zones not all of the concretions are fossiliferous. Geographical differences in the relative abundance of fossiliferous to non-fossiliferous concretions are obvious and there are also differences in the relative abundance of fossiliferous concretions from zone to zone, but these can only be assessed in qualitative terms because of the limited, mostly two-dimensional, exposures.

The distinction drawn here between assemblage zones and concretion layers characterized by a fossil name, like *Nucula* concretion layer or *Sphenodiscus* concretion layer, is largely one of relative abundance of fossils. In assemblage zones fossiliferous concretions are relatively common and each fossiliferous concretion usually has an abundance of fossils. In a concretion layer such as the persistent *Nucula* layer, concretions are generally barren and where fossils are present only one or two are commonly found in any single concretion. Disregarding the variety and abundance of fossils, both the assemblage zones and the concretionary layers with occasional fossils are characterized by a particular association of species, and one is as significant as the other in reflecting environment.

THE PIERRE-FOX HILLS CONTACT

Silt laminae and markedly siltier shale begin to appear in the upper part of the Elk Butte Member and locally become conspicuous. However, the very obvious increase in silt content that marks the beginning of the Fox Hills Formation is not likely to be mistaken anywhere in the type area. At a few places, gradation from the typical Elk Butte lithology takes place through as much as eight or ten feet of beds, but more commonly one or more very silty layers occur that can be used to mark the contact. Relative to key beds in the Lower Fox Hills the contact shows very little vertical variation throughout the type area and the total fluctuation nowhere appears to exceed about 15 feet; even this is excessive for most of the area and obtains chiefly in the Moreau River valley west of South Dakota Highway 63.

The silt layers and pods at the base of the Fox Hills are commonly impregnated with jarosite and the bright yellow accumulations of this mineral serve to accentuate the contact. Jarosite may also occur in the silty layers of the Elk Butte, but here it is much less concentrated and not persistent as at the base of the Fox Hills. In their original description of the Trail City Member, Morgan and Petsch (1945, p. 10) considered the bright yellow jarositic zone (which they refer to as "bentonite") to be in the top of the Pierre Shale and they placed their Pierre-Fox Hills contact between it and the first horizon of fossiliferous concretions. Subsequent usage has changed this practice so that the jarositic zone is now generally taken as the base of the Fox Hills.

The jarositic zone is only a secondary manifestation of the lithologic change from the silty shale of the Elk Butte to the clayey silt of the Lower Fox Hills and one can be easily misled by relying on it as primary indicator of the contact. Jarositic silt and sand occur at a number of horizons in the Lower Fox Hills and some of these are even more persistent than the basal jarositic zone.

The latter is more commonly present where the Little Eagle lithofacies of the Trail City Member overlies the Elk Butte Member of the Pierre Shale, although it also occurs intermittently at the base of the Irish Creek lithofacies. Awareness of the succession of the key beds in the Little Eagle lithofacies should eliminate the possibility of mistaking one of the higher jarositic zones for the basal one; in addition, both the silver-gray weathered color and the general absence of lamination or planar structure of any kind in its basal beds contrast strongly with the underlying Elk Butte Member.

Where the thin-bedded silt and silty shale of the Irish Creek lithofacies overlies the Elk Butte Member the contact is less obvious and the presence of more than one jarosite is a potential source of confusion. The difference between the weathered color of the two units is slight and the gradation from one lithology to the other is accentuated by the presence of planar structures on both sides of the contact. The basal jarosite is only present locally, but commonly the lower few feet of the Fox Hills is stained a dark rusty brown—possibly as a result of seepage along the thin beds of porous silt.

Although the Elk Butte shale and the lower part of the Irish Creek lithofacies are somewhat similar in their gross lithology the increase in silt content is pronounced and unmistakable; no difficulty was experienced in locating this contact all along the Moreau Valley to the point where it disappears beneath the surface in the vicinity of Irish Creek. Along the Grand River valley the contact is even more evident because the Little Eagle lithofacies is present along the river as far west as Bullhead where the contact passes beneath the surface.

TRAIL CITY MEMBER

DEFINITION AND TYPE SECTION

The name Trail City Member is given to the light-grey-weathering clayey silt that forms the lower 65 to 210 feet of the Fox Hills Formation in its type area. Morgan and Petsch (1945, p. 13-14) intially described the Trail City along exposures that extend from the vicinity of the town of Trail City on the east end of the Moreau-Grand divide, southwestward along the north side of the Moreau to St. Patrick's Butte (formerly Ragged Butte) just west of S. Dak. Highway 65. In all but the westernmost 2 or 3 miles of this outcrop area the Trail City beds are overlain by the Timber Lake sand, and throughout the eastern three quarters of it they contain the abundantly fossiliferous concretions that are its most conspicuous feature in its type locality around Trail City.

Measured sections presented by Morgan and Petsch (1945, p. 19-29) compared with remeasurements made during the present study indicate that they selected progressively higher horizons as the base of the Trail City in working westward along the Moreau River. They mistook the fossiliferous beds of the lower Timber Lake Member for the lower Trail City west of where the latter loses its rich fossil assemblages, and compounded the miscorrelation by locally mistaking both the medial and upper jarosites for the lower jarosite. Underlying beds equivalent to the Trail City, which become darker gray westward in the steeper, less-weathered, inner walls of the valley, superficially resemble the Pierre Shale and were mistaken for it.

Considering the lateral changes involved and the fact that the tracing was being done without the control of a detailed sequence of key beds the miscorrelation is understandable. It led Morgan and Petsch to conclude (1945, p. 13) that the Trail City Member becomes increasingly sandy to the west until "there is a sharp color change where the gray silty Pierre shale underlies the bright yellow ferruginous concretionary sand of the lowermost Fox Hills." They were followed in this by Curtiss (1952) in the only subsequent publication dealing with the stratigraphy along the Moreau west of typical exposures of the Trail City Member. Actually just the opposite is true; the Trail City lithology increases in thickness westward at the expense of the overlying Timber Lake Member, with gray, clayey silt replacing the yellowish sand. Beyond the western limit of the Timber Lake Member, which on the Moreau is between St. Patrick's Butte and S. Dak. Highway 65, the Trail City constitutes the entire Lower Fox Hills.

A case could be made for interpreting Morgan and Petsch as intending to restrict the name Trail City to the fossiliferous phase of the clayey silt. As will presently be shown, the abundantly fossiliferous concretion layers—although a striking lithologic feature of the typical Trail City—are distributed over an elongate, NNE-trending area that covers less than half the type area and terminates, on the south, within it. Because of this restricted distribution of supposedly typical features it is far less confusing and more practical stratigraphic procedure to define the Trail City Member on its overall lithology and include in it all of the clayey silt of the Lower Fox Hills. For environmental stratigraphy, however, the separate facies need separate recognition; the informal terms Little Eagle lithofacies and Irish Creek lithofacies are introduced, respectively, for the fossiliferous, so-called "typical", Trail City and for the sparsely fossiliferous Trail City into which both the Little Eagle lithofacies and the lower part of the Timber Lake Member grade.

The details of Trail City stratigraphy are presented below in the separate descriptions of its two lithofacies, for each of which reference sections are given. No type section was designated for the Trail City Member by Morgan and Petsch (1945, p. 13); the "excellent development in the area of Trail City," for which it was named, presently lacks exposures from which an adequate type might be selected. None of five selected reference sections given by Morgan and Petsch (1945, p. 19-28) to illustrate their Pierre-Fox Hills sequence is in the Trail City area, but their section 4 (ibid, p. 24-25) and its adjacent outcrops contain the only easily accessible, complete section of the Trail City in the entire type area of the Fox Hills Formation. This locality, 30 miles airline southwest of the town of Trail City along S. Dak. Highway 63 just north of the Moreau River, includes numerous exposures which afford a composite section from within the Elk Butte Member of the Pierre Shale well into the Timber Lake Member of the Fox Hills. Here the Trail City Member has begun its lateral change from the Little Eagle to the Irish Creek lithofacies and the fossils are not as abundant as they are a short distance to the east; the succession shows most of the common key beds and with a little search representative fossils can be collected from the assemblage zones. Section 1 (p. 81), measured at this locality, constitutes the best available reference section for the Trail City Member. Although it is essentially the type section by subsequent designation, strict interpretation of the Code (Am. Comm. Strat. Nomenclature, 1961, Art. 13a, h, 1) apparently would only permit calling it the principal reference section because it does not lie within the originally designated type locality. Sections 2 through 7 include type and reference sections for the two lithofacies of the Trail City and consequently are reference sections for the member as a whole.

The Trail City Member as defined here extends throughout the type area of the Fox Hills Formation lying between the Pierre Shale and Timber Lake Member in the eastern part and between the Pierre and the Iron Lightning Member in the western part. It remains to be determined how far to the east and northeast beyond the type area of the Fox Hills the member can be usefully distinguished. It is in these directions, and not to the west and southwest, that the member becomes increasingly sandy and grades in large part into the Timber Lake lithology.

The western limit of the Fox Hills type area is the west line of Ziebach County; the Trail City Member, here in the Irish Creek lithofacies, consequently extends to where the county line intersects its single band of outcrop qn the south side of the Cheyenne-Moreau divide. I suspect this will prove to be a propitious place to terminate arbitrarily the Trail City Member, as the Irish Creek lithofacies begins to change beyond this point, becoming increasingly sandy. However, no decision on this is warranted until the area west of Ziebach County is thoroughly studied; to date the Irish Creek equivalent there has been included in the Pierre Shale.

The persistence with which geologists have assigned the clayey silts of the Trail City to the Pierre Shale is evident in the historical survey of the type Fox Hills and cannot be ignored here. In large part it has been due to the lack of a rigorous lithologic definition of the Pierre-Fox Hills contact and the tendency to rely solely on the most superficial features of the outcrop. This is substantiated by the fact that the basal clayey silts of the Lower Fox Hills have been recognized as part of that formation in all the detailed quadrangle studies of the type area by the South Dakota Geological Survey, whereas, from the work of Stanton (1910) to that of Tychsen and Vorhis (1955), it has been chiefly in reconnaissance work and economic studies not concerned with the stratigraphy of the marine beds that the clayey silts have been considered Pierre Shale.

The problem is more than a matter of nomenclatural semantics; a logical, preferably convenient, line of demarcation must be selected in an admittedly transitional sequence. In the type area this poses two questions: 1) can the Trail City clayey silt be separated consistently from the underlying Elk Butte Member, and 2) if so, why should it be considered a basal phase of the Fox Hills rather than part of the Pierre Shale? The facts already brought out in the description of the Pierre-Fox Hills contact leave no question that throughout the area a lithologic separation based on the relatively sharp increase in quantity and grain size of silt and/or very fine-grained sand can be made between the lower clayey part of the Fox Hills and the Pierre Shale.

Ample justification for including the clayey silts in the Fox Hills Formation is found in their close stratigraphic relationship with the sand lithofacies (Timber Lake Member) in the Lower Fox Hills. Not only do they grade laterally into it, but they share common features such as the layers of fossiliferous limestone concretions that are quite different from anything in the underlying Elk Butte Member. But perhaps the strongest justification is that the base of the clayey silts marks a change from uniformity of deposition over a wide area (characteristic of off-shore marine deposits) to diversity of deposition within limited areas (characteristic of marginal marine environments). The contact at the base of the clayey silt has validity both as a lithologic and lithogenetic boundary. The alternative choice of the base of the first sand in the sequence is too crude a differentiation to be useful in detailed stratigraphic work; it has only local lithogenetic significance and use of it as the base of the type Fox Hills, within which the sand body pinches out completely, would obscure significant stratigraphic relationships and make for an unwieldy subdivision.

LITTLE EAGLE LITHOFACIES

The Little Eagle lithofacies of the Trail City Member is named for exposures in the breaks on the north side of the Grand River north and west of the village of Little Eagle, Corson County. The type section, located on a south-facing bluff approximately 4 miles WSW of the village is described in Section 3, p. 84. Sections 1 and 2 are additional reference sections of the lithofacies.

The most distinctive features of the Little Eagle lithofacies are the abundantly fossiliferous concretion layers in its lower half and the mixed nature of its sediment, which was so thoroughly worked by burrowing organisms that bedding structure is largely obliterated. Separate recognition of this lithofacies serves to define, geographically, a portion of the Trail City Member that includes an exceptional fossil record useful in the environmental and ecological interpretation of the Fox Hills Formation.

DISTRIBUTION

The Little Eagle lithofacies occupies the interval between the Pierre Shale and the Timber Lake Member of the Lower Fox Hills over most of the eastern part of the type area. Its principal areas of outcrop are around the edges of the eastern part of the Moreau-Grand divide and along the bluffs on the north side of the Grand River. Westward along the north side of the Moreau Valley it begins to lose its fossil accumulations rather abruptly in the vicinity of South Dakota Highway 63 in the west half of the Parade NW quadrangle of the U.S.G.S. 7½-minute topographic series. However, sporadic fossil accumulations occur westward into the adjoining east half of the Lantry N.E. quadrangle with sufficient regularity to permit recognition of the lithofacies; long 101°15'W., along which these two quadrangles join, affords an arbitrary western limit for the Little Eagle lithofacies in the Moreau Valley.

On the south side of the Moreau, fossiliferous concretions occur near the river between Green Grass village, approximately long 101°15'W., and the longitude of Whitehorse village about 22 miles to the east-northeast. They become increasingly scarce where tributaries to the Moreau, such as Goose Creek, can be followed southward into the Cheyenne-Moreau divide. On the south side of the Cheyenne-Moreau divide good exposures of the Lower Fox Hills beneath the Timber Lake Member are rare. In the only two partial exposures examined, one south of Eagle Butte, the other southwest of Ridgeview, the Trail City Member was in the Irish Creek lithofacies and completely barren of fossils. Although two outcrops are insufficient number for saying that the Little Eagle lithofacies nowhere extends to the south side of the Cheyenne-Moreau divide, this is strongly indicated by the general lack of its characteristic fossiliferous concretions in the soil and slopewash along its projected outcrop area.

In the Grand River valley the Little Eagle lithofacies passes into the Irish Creek lithofacies somewhat west of where its basal contact drops below river level, a few miles southwest of Bullhead in the northeast corner of the Miscol quadrangle of the U.S.G.S. 7½-minute topographic series, approximately within sections 32, 33, and 34. T.21N., R.24E. Here, as in the Moreau Valley, a few scattered fossiliferous concretions are found beyond the last exposures in which they are abundant, but the change is definitely more abrupt and takes place within a few miles. The change here is considerably closer to long 101°07'30"W. than to long 101°15'W. To the east the Little Eagle lithofacies can be traced by means of its characteristic fossil assemblages along the north side of the Grand River as far as U.S. Highway 12, where the layers begin to become less fossiliferous. East of here the outcrop of the Trail City Member in the bluffs south of Rattlesnake Butte is largely obscured by slump or overgrown, but partial exposures and scarcity of characteristic fossil associations in the float indicate a change from the Little Eagle lithofacies.

The valley of Oak Creek, northeast of the Grand River, marks the approximate limit of the type area of the Fox Hills and the formation was not studied in detail beyond it. Work by Speden (1965, pls. 2-13 and measured sections from field notes) along and northeast of Oak Creek indicates a marked change in the Trail City and substantiates that the Little Eagle lithofacies does not extend northeastward across it in this area. However, the lower two assemblage zones of the Little Eagle lithofacies occur in the base of the Fox Hills Formation 40 miles airline northeast of Oak Creek in the vicinity of Linton, North Dakota. Here they occur in 24 feet of Trail City lithology between the Pierre Shale and the Timber Lake Member. Either this thin representative of the Little Eagle lithofacies is a distinct body from that in the type area or the eastern margin of the latter extends north from the Grand River, under cover of younger rocks, into North Dakota before turning eastward across the Missouri Valley.

Throughout its area of outcrop the base of the Little Eagle lithofacies generally shows the lower jarositic silt zone. Even where this key bed is locally absent the approximate base is indicated on weathered outcrops by the sharp color change from gray to light gray and by the abundant gray-weathering concretions in the Lower *nicolleti* Assemblage Zone just above the contact.

The sand of the Timber Lake Member commonly begins to appear about at the horizon of the upper jarositic silt zone (see Fig. 7). This key bed is locally useful both in locating the upper contact and in serving as the contact where the change from the clayey sand and silt of the upper Little Eagle is gradual through an appreciable interval. In the southern outcrops of the lithofacies the sand becomes the dominant constituent from 10 to 25 feet above the upper jarositic silt whereas in the northeastern outcrops—east of the Bullhead area—the sand begins 10 or 15 feet below the horizon of this key bed. The upper jarositic silt is not present where its position is occupied by sand, but locally one or two thin beds of highly glauconitic sand mark its position.

Using the upper jarositic silt as the upper contact, the Little Eagle lithofacies of the Trail City Member ranges between 65 and 100 feet in thickness; using the actual base of the Timber Lake sand the range is greater but the average thickness is still about 80 feet.

LITHOLOGY

Clayey silt and clayey sand are the predominant types of sediment in the Little Eagle lithofacies. These sediments are very rarely and only locally homogeneous in structure. The mixing of the clay and coarser fraction is incomplete and is characterized by blebs and stringers of silt or clay. The structure is identical to that of recent sediments that have been worked by organisms. Local thin-bedded zones and lenses are fairly common and show that the Trail City sediment was deposited as alternating layers of silty shale and silt or sand, the latter commonly irregular in thickness and displaying internal lamination and cross-lamination.

In addition to the "worked" clayey sediments the Little Eagle lithofacies contains zones of sand, silt and, less commonly, silty shale. The word zone rather than bed is used because these units are not sharply bounded, but generally grade into adjacent sediments. Bedding planes are uncommon in much of the lithofacies.

The clay fraction of the Little Eagle is locally as high as 70 percent and probably averages between 55 and 60 percent for the lithofacies as a whole. But such generalized quantitative data can be misleading locally because of the considerable variation from place to place. In general, the clay content is sufficient to give the lithofacies the aspects and weathering properties of a dominantly clayey unit. The clay itself is gray to dark gray in fresh exposures.

The coarse fraction of the Little Eagle ranges from silt to medium-grained sand. The silt blebs, lenses and laminae weather white, but are commonly stained brown and rusty brown or impregnated with yellow jarosite. The sandy layers are commonly glauconitic and almost invariably weather brown or reddish brown; locally the glauconite is concentrated enough to give thin beds a greenish cast.

The lithofacies as a whole weathers to a step-like outcrop, individual concretion layers forming the local benches. Exposures range in color from light silvery-gray to yellowish and brownish-gray; light gray dominates the lower part and the yellow and brown tints appear chiefly on middle and upper parts of the lithofacies. Local concentrations of jarosite and iron oxides result in color bands of light values of gray, brown, and yellow grey, on some exposures. On close inspection the typical Little Eagle outcrop is mottled, reflecting the irregular mixture of clay with the silt and sand.

The Trail City Member has been referred to repeatedly as "bentonitic", but this is misleading. The sediment does contain shards in its finer clastic fraction and the Irish Creek lithofacies has a few thin bentonite beds. But in neither the latter nor the Little Eagle lithofacies is bentonitic material more than a minor constituent. In the Little Eagle the reworking of the sediment has dissipated actual bentonite layers so that material that can properly be called bentonitic is uncommon. The confusion of bentonite with jarosite, which is widespread in both lithofacies, has been fairly common in reports on the type area of the Fox Hills and is probably the source of the errors.

Sections 1 through 3 (p. 81-86) illustrate the principal features of the Little Eagle lithofacies. Although local variations in parts of this sequence of beds are common, they are usually variations in degree rather than kind. The lower half of the lithofacies, which contains most of the fossiliferous concretion layers, is chiefly clayey silt, but in and just above the *Protocardia-Oxytoma* Assemblage Zone, this locally changes to clayey sand or fine-grained, commonly glauconitic sandstone. This sandy interval is thicker and less clayey in the outcrops of the eastern part of the Grand River valley and the eastern part of the Moreau-Grand divide, where it commonly is highly glauconitic. Westward in both river valleys it is less conspicuous and the glauconitic sandy beds above the *Protocardia-Oxytoma* concretions grade into silt and clayey silt with abundant jarosite—the medial jarositic silt zone. Distribution of the sand is shown in fig. 19.

The few feet just above the medial jarositic silt, and its lateral sandy phase, is the only part of the Little Eagle lithofacies in which shale is commonly present. Here may be found as much as 10-12 feet of finely silty, dark-gray shale, locally weathering to a checked surface with a brownish-gray stain. The locally abrupt contact of the shale with the underlying silt or sand is one of the few sharply defined bedding surfaces in the sequence. Upward, the shale grades back into clayey silt or to clayey sand before the persistant *Nucula* concretion layer is reached. Between the *Nucula* concretion layer and the upper jarositic silt zone the upper part of the lithofacies varies chiefly in the relative amount of clayey silt and clayey sand it contains; in general, the sand content increases to the east and northeast.

CONCRETION LAYERS AND ASSEMBLAGE ZONES

Description of the successive concretion layers and zones of the Little Eagle lithofacies should be prefaced by the reminder that in spite of their local variations in character and distribution the most striking feature of these layers is their overall uniformity in character and stratigraphic position throughout their extent.

Lower *nicolleti* Assemblage Zone: The base of the Little Eagle lithofacies contains abundant silty limestone concretions with thick silt jackets, some of which are fossiliferous and characterized by large numbers of the ammonite *Scaphites* (*Hoploscaphites*) *nicolleti* (Morton) and a less conspicuous fauna of other molluscs. The concretion zone persists without known break throughout the area of the lithofacies, although it is not everywhere abundantly fossiliferous. It either recurs or extends as far to the northeast as the vicinity of Linton, in Emmons County, North Dakota, where it is abundantly fossiliferous. Distributional features of the zone in the type area of the Fox Hills are shown in Fig. 16, and typical fossiliferous concretions are shown in Plate 2.

The interval occupied by concretions ranges from 2 to 15 feet in thickness. In some places concretions begin within a foot of the base of the Little Eagle, in others as much as 10 feet above the base; correspondingly, the top of the zone may be anywhere from 2 to 25 feet above the base of the lithofacies. The variation in distance above the base may result from local variations in the thickness of sediment deposited but it may also reflect local difficulty in placing the contact where the change across it is more gradational than usual or the basal jarositic silt is poorly defined or absent.

Variations in the thickness of the Lower *nicolleti* Assemblage Zone are commonly associated with lateral changes in its constitution. The zone is complex and exhibits two major phases (see Fig. 16): a thinner eastern phase usually characterized by several layers of large concretions that here and there include an abundantly fossiliferous one, and a somewhat thicker western phase characterized by numerous, small (1.5 to 3 inches diam.), spherical concretions scattered among the larger kind. The abundant *nicolleti* accumulations disappear westward within the latter phase, where fossils tend to occur only in some of the small concretions, usually as single specimens.

The concretions in the Lower *nicolleti* Assemblage Zone are distinctive in their lithology and can be distinguished from those in overlying layers and assemblage zones on this characteristic alone. Most have spheroidal to flat-ovoid cores of gray, silty limestone from 6 to 18 inches in diameter; these are enveloped in a tough jacket of calcareous silt or very fine sand that weathers light gray and locally displays lamination and cross-lamination. Commonly the cores are little more indurated by calcium carbonate than the jackets and are punky rather than brittle; locally, no well-defined core is evident. Where several layers of concretions are present in the Lower *nicolleti* zone, the cores of successive layers become less silty, harder and darker gray upward as they become silty limestone rather than calcareous siltstone. The cores of highly fossiliferous concretions are generally silty limestone in whichever layer they occur.

The jackets of the Lower *nicolleti* concretions are distinctive for several reasons. They weather light gray and are quite thick; at a few places where concretions are closely spaced they coalesce to form a crumbly ledge. Locally the jackets show bedding structure consisting chiefly of thin, irregular, cross-laminated sets. In the concretions with highly silty cores, bedding structures can commonly be traced from the jacket into the core without a break. Bedding structure is limited to local areas where the sediments have not been thoroughly mixed by organisms; it is more common in the lowest layer, or layers, of concretions in the Lower *nicolleti* zone. Many of the concretions have an outer rind, half an inch or less in thickness, of porous gypsum impregnated with a black stain. This black rind appears to be a unique feature of the Lower *nicolleti* concretions; it has not yet been found in higher concretion layers. Most of the fossiliferous concretions collected in place had the black gypsiferous rind, and in prospecting for concretions not yet exposed the patches of black gypsum fragments proved a useful guide.

In the western phase of the Lower *nicolleti* Assemblage Zone these typical concretions become interbedded with the smaller, generally jacketless, silty limestone concretions and as these increase in number the larger, jacketed concretions decrease in number. The change from one phase to the other is gradual and rarely complete. A few small concretions are commonly found scattered among the larger ones in the eastern phase; here they are generally unfossiliferous or contain single large specimens of *Gervillia*. In the western phase the smaller concretions may be sparsely scattered or abundant and may or may not include an occasional larger, jacketed concretion; the local variation is great and the only consistent feature is the lack of accumulations of *nicolleti* and the presence of a few small concretions bearing single specimens of this and one or two other scaphites.

Certain local variations in fossil associations within the Lower *nicolleti* Assemblage Zone indicate a more complex distribution pattern among molluscs other than *S. (H.) nicolleti*. One obvious variation found locally in the uppermost concretion layers of the Lower *nicolleti* zone is the occurrence of relatively small concretions dominated by the gastropod *Drepanochilus*. These accumulations occur in limited areas of outcrop, in both the Moreau and Grand valleys, on the west side of the area covered by the *nicolleti* assemblages (Fig. 16). To the east and partially overlapping the *Drepanochilus* areas of accumulation large specimens of *Gervillia* are noticeably more abundant than elsewhere in the *nicolleti* zone. These relatively conspicuous distributional features noticed in the biostratigraphic field work were born out by Speden's (1965) subsequent detailed work on the bivalves, which revealed still more complex associational features.

Limopsis-Gervillia Assemblage Zone: From 13 to 30 feet above the base of the Little

Eagle lithofacies a persistent layer of concretions occurs that in some places constitutes the only layer, and in others the basal one of several layers, in the *Limopsis-Gervillia* Assemblage Zone. The basal layer generally lies between 5 and 10 feet above the uppermost concretion layer in the Lower *nicolleti* Assemblage Zone, but locally may occur as little as 2 feet and as much as 13 feet above it; the intervening sediments lack key beds or any persistent distinctive features. Distribution of the *Limopsis-Gervillia* zone is shown in Fig. 17, and examples of its fossiliferous concretions in plate 2.

The *Limopsis-Gervillia* zone consists of only the basal concretion layer throughout most of the Moreau River valley outcrop of the Little Eagle lithofacies. Here the concretions consist of hard, bluish-gray limestone that weathers a rich, rusty-brown color. Scattered concretions are crowded with the small infaunal bivalve *Limopsis striatopunctatus*, and an assortment of much less numerous molluscan species, the more common of which are the epifaunal bivalve *Gervillia subtortuosa* Meek and Hayden and a small sulcate bivalve which Meek called *Nemodon sulcatinus*. Ammonoids, though relatively few in numbers, are rich in variety and include small specimens of the genus *Sphenodiscus* and, in addition to the endemic scaphite species found in the Lower *nicolleti* Assemblage Zone, a variety of immigrants from the Gulf Coast region including *Discoscaphites conradi* (Morton) (not Meek), and the tiny *Baculites columna* Morton.

The localities along the north side of the Moreau River east of the road to Whitehorse contain an occasional concretion crowded with small specimens of *Gervillia* in the beds from 2 to 5 feet above the brown-weathering *Limopsis*-bearing concretion layer. In the Grand River valley area, the *Gervillia*-bearing concretions are common and form a persistent part of the *Limopsis-Gervillia* Assemblage Zone. Here they occur in one to three separate layers, usually beginning about 2 feet above the basal layer which generally has the same characteristics and fauna as in the Moreau Valley. Locally, however, the basal *Limopsis* layer contains concretions with masses of very large specimens of *Gervillia*. Where the *Gervillia*-bearing concretions are present the *Limopsis-Gervillia* zone varies between 3 and 10 feet in thickness.

The *Gervillia*-bearing concretion layers are the only ones in the entire Little Eagle lithofacies in which the fossiliferous concretions appear to outnumber the unfossiliferous. The concretions are ovoid and flat-ovoid, usually lack jackets, and consist of bluish-gray limestone that weathers gray or brownish gray. However, it is the basal layer with the scattered, *Limopsis*-bearing concretions that is the most distinctive and persistent layer of the assemblage zone. It contrasts markedly with the underlying Lower *nicolleti* concretions in the rusty-brown weathered color, which locally also stains the adjacent sediments.

Locally, in the Grand River valley 3 miles south of Bullhead, barrel-shaped concretions with vertical long axis include the *Limopsis* association in the lower part and the *Gervillia* association in the upper part. Here the entire assemblage zone, which consists of these faunally compound concretions and a second layer of *Gervillia*-bearing concretions just above it, is no more than 3 feet thick. Additional telescoping of the section is seen in the overlying beds where only 5 feet of sediment separate the upper layer of *Gervillia* concretions from the concretions of the *Protocardia-Oxytoma* Assemblage Zone; the Upper *nicolleti* Assemblage Zone is missing. Evidently there was considerable by-passing of sediment in this local area during the deposition of the lower part of the Little Eagle lithofacies for there are no obvious breaks in the section. But this does not explain the juxtaposition of the *Gervillia* clusters on the *Limopsis* clusters to form single masses. Any physical means of bringing these masses together involves unbelievable coincidence. More likely the *Limopsis* accumulation, partially exposed by currents, furnished a place of attachment for clusters of the epifaunal, byssate *Gervillia*.

Because of the astronomical numbers of *Gervillia* and *Limopsis* shells in the assemblages it is the easiest zone to locate and trace where the Little Eagle outcrop is grassed over. Geographically, the *Limopsis-Gervillia* Assemblage Zone is as widespread as the Lower *nicolleti* Assemblage Zone; its fossil accumulations disappear at approximately the same places to the west, and they are found as far northeastward as the Linton area of North Dakota. Together, these two assemblage zones are the most widespread in the Fox Hills Formation.

Upper *nicolleti* Assemblage Zone: Between the *Limopsis-Gervillia* Assemblage Zone and the next persistent fossiliferous concretions (those of the *Protocardia-Oxytoma* Assemblage Zone) lies a variable part of the Trail City Member ranging from 5 to 30 feet in thickness. In some places, particularly where it is under 15 feet thick, this interval is barren of concretions, but throughout a large part of the eastern half of the type area it contains a zone of concretions with sporadic accumulations of fossils dominated by *Scaphites* (*Hoploscaphites*) *nicolleti*. In addition, a small area in and around the southwestern limit of the Little Eagle lithofacies on the Moreau River has in this interval scattered, small, sparsely fossiliferous concretions containing occasional ammonoids, pelecypods, and less commonly gladii of the coleoid cephalopod *Actinosepia*.

These two fossil accumulations, here called the Upper *nicolleti* Assemblage Zone and *Actinosepia* concretions, respectively, occupy approximately the same stratigraphic position—an interval from 2 to 12 feet below the concretions of the *Protocardia-Oxytoma* Assemblage Zone. They appear to bear a relationship to one another somewhat similar to that between the typical phase of the Lower *nicolleti* Assemblage Zone and its western phase with the small concretions. However, the *Actinosepia* concretions rarely occur at localities where the larger Upper *nicolleti* accumulations are present. The total faunas of each are very similar to one another and to the far richer, overlying *Protocardia-Oxytoma* assemblages, but there is a marked difference in dominant species.

The Upper *nicolleti* Assemblage Zone consists of one or locally two rather poorly defined layers of hard, generally ovoid, dark blue-gray limestone concretions. Individual concretions are rather widely spaced throughout most of the area in which the zone occurs (Fig. 18), being from 10 to as much as 30 feet apart on most outcrops; at a few places they are more closely spaced. Sporadic fossiliferous concretions consist mostly of *S.* (*H.*) *nicolleti*, with other scaphites including a coarsely spinose form similar to *Discoscaphites conradi* var. *navicularis* (Morton). Associated pelecypods are most commonly dominated by "*Inoceramus*" *fibrosus*. The *nicolleti-fibrosus* association is also a common association in the Lower *nicolleti* Assemblage Zone.

The Lower and Upper *nicolleti* fossil assemblages can be distinguished by the presence of numerous ammonites of the *conradi* complex in the latter and generally by the more varied associated bivalve fauna in the former. The concretions themselves are also quite distinct in character; those of the Upper *nicolleti* zone generally lack jackets and are harder, darker, less silty limestone. Nevertheless, the fossil associations of the two horizons are remarkably similar and in dealing with concretions in float it can be very difficult to tell them apart, particularly if *fibrosus* is the dominant pelecypod and no ammonites other than *nicolleti* are present.

The *Actinosepia* concretions are chiefly barren, brittle, blue-gray limestone and vary in shape from spheroid to ovoid and flat ovoid. They are generally 8 inches or less in long diameter but rarely smaller than 3 inches. Fossils occur sparingly; scaphites, the most common, occur singly and are chiefly large specimens of *S. (H.) nicolleti* or a similar-looking but slightly more nodose species resembling Meek's *Discoscaphites conradi* var. *intermedius*. Gladii of *Actinosepia* are not common relative to ammonoids; they are rare fossils and only seem common because the concretions in this small area of Trail City outcrop along the Moreau have yielded almost twice the number previously reported from the Cretaceous of all of North America. Other fossils found rarely in these concretions include small lobster-like crustaceans, local accumulations of the bivalve *Oxytoma nebrascana* (Evans & Shumard), and rare stems of *Palmoxylon* up to 8 feet in length (Delevoryas, 1964). Additional details on *Actinosepia* and its distributional features are published elsewhere (Waage, 1965).

The distribution of the Upper *nicolleti* Assemblage Zone and the *Actinosepia* concretions is shown in Fig. 18. The latter extend well into the Irish Creek lithofacies westward along the Moreau River, but have not been found in other areas of Irish Creek exposures. Richly fossiliferous Upper *nicolleti* concretions are present at least as far north as the breaks of Little Oak Creek and adjoining bluffs south of the Grand River, near Little Eagle, but none have yet been found either north of the Grand River, or west along the Grand from this area. Westward from the village of Little Eagle the thickness of the interval between the uppermost concretions of the *Limopsis-Gervillia* zone and the basal ones of the *Protocardia-Oxytoma* zone drops appreciably. At locality 50, 4 miles west of the village, it is only 10 feet thick, and westward from here as far as the zones can be traced it is between 5 and 10 feet thick. Throughout this area where the interval is unusually thin it generally lacks concretions of any kind. The coincidence of a thin interval with the absence of the Upper *nicolleti* concretions suggests lack of sedimentation in the area at the time the Upper *nicolleti* or equivalent accumulations were formed. A comparable but less drastic thinning was noted in the same area within the underlying *Limopsis-Gervillia* zone in which occur the unusual vertically elongated concretions with *Limopsis* accumulations in the lower half and *Gervillia* accumulations in the upper half. Convergence of the Upper *nicolleti*, or equivalent, accumulations with those of the *Protocardia-Oxytoma* Assemblage Zone due to the drop in sedimentation rate is also a possibility, but it would be difficult to detect, as most species found in the Upper *nicolleti* zone are also common in the *Protocardia-Oxytoma* zone. Nevertheless it should be possible to distinguish assemblages of the two layers because the *Protocardia-Oxytoma* zone, though it commonly contains concretions made up largely of ammonoids, lacks concretions dominated by the *S. (H.) nicolleti* association. Unless more definite evidence can be found it can only be assumed that neither the Upper *nicolleti* nor *Actinosepia* assemblages accumulated in the area in question.

The apparent absence of Upper *nicolleti* concretions north and northeast of Little Eagle may be an artifact of the distribution of good exposures, but it seems unlikely that the highly fossiliferous concretions from this layer could have gone undetected even on grass-covered slopes. The interval between the *Limopsis-Gervillia* and *Protocardia-Oxytoma* zones is just as thick here as in the richly fossiliferous area south of the Grand River, so there is no reason to suspect omission due to lack of sedimentation. Until future collecting proves to the contrary, the Grand River must be assumed to mark the approximate northern limit of the Upper *nicolleti* accumulation.

Protocardia-Oxytoma Assemblage Zone: The first appearance of an appreciable amount of sand in the Little Eagle lithofacies is associated with the concretions of the *Protocardia-Oxytoma* Assemblage Zone. At many places the typical clayey silt grades upward into very fine-grained to medium-grained, clayey, somewhat glauconitic sand 4 or 5 feet below the first concretions in the zone and continues for as many feet above the zone. The sand is localized in distribution, occupying a north-northeast-trending area on the eastern parts of the divides (Fig. 19). The zone of concretions varies from a single layer to an interval with scattered concretions as much as 10 feet thick; at most localities it ranges from 1 to 5 feet in thickness.

The *Protocardia-Oxytoma* zone lies approximately in the middle of the Little Eagle, varying in position from 30 to 55 feet above the base; it marks the top of the abundantly fossiliferous lower part of the lithofacies. The marked variation in its position relative to the *Limopsis-Gervillia* zone was discussed in the preceding section on the Upper *nicolleti* zone; the interval is generally thicker along the Moreau River (from 15 to 30 feet) than along the Grand River (from 5 to 20 feet).

The concretions in the zone are hard, dark blue-gray limestone; in sandy beds they weather reddish brown and generally are surrounded by semi-indurated jackets that may coalesce to form a crumbly, sandy ledge. Where the matrix is dominantly clayey, jackets are less common and the limestone cores weather light brownish-gray or have a thin off-white or light-gray patina. Most of the cores are between 6 and 15 inches in diameter and vary in shape from spherical to flat-ovoid.

Considerable numbers of two bivalves, the infaunal *Protocardia subquadrata* (Evans and Shumard) and epifaunal *Oxytoma nebrascana,* characterize the fossiliferous concretions, which also have a plentiful assortment of other fossils. These two species commonly occur in about equal quantity, each far outnumbering the individuals of other species in the fauna. Their usual mode of occurrence is in separate clusters of one species or the other; a single concretion may have a mass of *Protocardia* in one part and a mass of *Oxytoma* in another. The epifaunal *Oxytoma* seems to be the more gregarious and commonly monopolizes individual concretions. Ammonoids are a plentiful element of the fauna in the *Protocardia-Oxytoma* zone, and concretions in which they are a numerically significant part of the assemblage are fairly common. Plant material ranging from fragments to pieces of wood and foliage is more common than in underlying concretion layers. In bluffs north of the Little Moreau River (Locality 95) a second calcified, stem-like fragment of *Palmoxylon* about 6 feet in length was found associated with, but not contained in, concretions of the *Protocardia-Oxytoma* layer.

As in the assemblage zones below it, the concretions of the *Protocardia-Oxytoma* zone become unfossiliferous westward. Generally the zone is reduced to a single layer of concretions near the western edge of the Little Eagle lithofacies and in the vicinity of South Dakota Highway 63 on the Moreau River, the typical fossil assemblages give way to sparsely fossiliferous concretions in which the pelecypod *Lucina* is locally a conspicuous member of the assemblage. The latter occurs in the concretions and also in the surrounding sediment where it is commonly found in living position. *Lucina* also occurs in abundance locally along High Bank Creek south of the Grand River, where the *Protocardia-Oxytoma* concretions lack their characteristic fossils. Apparently the association with *Lucina* is fairly common in marginal areas to the west of the typical *Protocardia-Oxytoma* assemblages.

The extent of the *Protocardia-Oxytoma* Assemblage Zone beyond the type area of the Fox Hills is not known. It is not present in the Linton area of Emmons County, North Dakota, so its lateral distribution is less than that of the Lower *nicolleti* and *Limopsis-Gervillia* zones. The identification of the *Protocardia-Oxytoma* Assemblage Zone within the type area is facilitated by the abundantly and distinctively fossiliferous concretions, the sandy beds in which they occur, and the widespread medial jarositic silt that overlies them; Fig. 18 shows its distribution, and typical concretions are illustrated on Pl. 2.

Upper Little Eagle Concretion Layers: The upper part of the Little Eagle lithofacies contrasts markedly with the lower part in the scarcity of fossiliferous concretion zones and its greater lateral variability. Although well-defined successions of concretion layers are discernible, it is locally difficult to trace a succession because of abrupt changes in thickness and lithology of the sediments between the concretion layers. Additional confusion is introduced by the presence of local concretion layers or scattered concretions. In spite of the variability in thickness between concretion layers, which may be as much as 20 feet within 6 to 8 miles, the entire interval between the basal concretions of the *Protocardia-Oxytoma* Assemblage Zone and the upper jarositic silt is rather consistently between 40 and 50 feet thick.

The upper jarositic silt is only intermittently present in exposures along the Grand River east of Bullhead, but elsewhere in the area of Little Eagle outcrop it is remarkably persistent. As previously noted, its absence in the eastern part of the Grand River valley is largely due to the lateral replacement of the uppermost Little Eagle lithofacies by the Timber Lake sand.

The most persistent key bed in the upper part of the Little Eagle lithofacies is the *Nucula* concretion layer which usually lies between 10 and 20 feet below the top. This layer consists of spherical to ovoid concretions ranging from a foot to as much as 2.5 feet in diameter and commonly spaced from 10 to 30 feet apart along the outcrop. The concretions are brittle, dark-gray limestone with yellow calcite filling radial and concentric fractures; they generally weather reddish brown to rusty brown and locally are encased in tough silty or sandy jackets. Most concretions are completely barren of fossils, but some contain one or two specimens. The bivalves *Nucula* and *Nuculana* are by far the most common fossils in this layer, although a few broken ammonoid shells, chiefly of *Discoscaphites nebrascensis* have been found. The fossils, though everywhere sparsely distributed, are more common in the Grand River area than along the Moreau River.

Most of the stratigraphic variation in the upper part of the Little Eagle lithofacies occurs between the *Nucula* concretion layer and the *Protocardia-Oxytoma* Assemblage Zone, some 20 to 40 feet beneath it. The sand body in and just above the *Protocardia-Oxytoma* Assemblage Zone along the eastern side of the type area (p. 113 and Fig. 19) is responsible for much of this variation. At its greatest stratigraphic extent this sand ranges from a few feet below the *Protocardia-Oxytoma* Assemblage Zone to the layer of barren concretions called the barren A layer (Fig. 7) which lies from as little as 7 to as much as 32 feet above the uppermost *Protocardia-Oxytoma* concretions. The barren A concretion layer is most useful in projecting the approximate top of the sand interval laterally into areas of clayey silt and shale peripheral to the sand body on the west. When this is done it is found that the greatest thickness of the interval lies off the west side of the sand body along a northeast-trending axis lying approximately between Little Eagle and the bridge across the Moreau River on S. Dak. Highway 63, roughly paralleling the axis of the sand body.

In the thicker parts of the sand body, which are chiefly on its western side, and in the shaly area of greatest thickness lying west of it, one or more local concretion layers are present that apparently do not occur elsewhere. Most of the latter are gray-weathering, brittle, unfossiliferous limestone concretions but in the sandier beds large, rusty-brown weathering, sandy limestone concretions are locally common. Rarely these concretions carry a few scattered fossils. The composition of these sparse faunules, which include small clusters of *Oxytoma,* relates them to the *Protocardia-Oxytoma* assemblages.

The *Nucula* layer, the barren A layer, and the local concretion layers below them are physically similar. They can be identified only by their relative positions between the medial and upper jarositic zones. Even though fossils have been found so far only in the *Nucula* layer, it would not be surprising if they also occur in the lithologically similar concretions of the barren A and other more local layers.

The only abundantly fossiliferous concretions in the upper part of the Little Eagle lithofacies are found scattered in the interval between the *Nucula* layer and the upper jarositic silt throughout most of the Little Eagle outcrop in the Grand River valley. These are chiefly small, jacketless, dark-brown to yellow-brown-weathering limestone concretions 10 inches or less in diameter that are crowded with a variety of fossils the most conspicuous of which are immature scaphites, mature specimens of *Discoscaphites abyssinus* (Morton), *D. mandanensis, D. cheyennensis* (Owen) and the bivalve *Pteria linguaeformis* (Evans and Shumard). The abundance of juvenile scaphitids is perhaps the most characteristic feature of these assemblages; juveniles are relatively rare elsewhere in the Trail City Member. The assemblages differ from those in the other Little Eagle assemblage zones in lacking the characteristic great abundance of one or two species and in containing fewer intact specimens; both features suggest more working by currents than is typical for most of the fossiliferous concretions. In these features and in the constitution of the fauna itself these assemblages are more like those of the Timber Lake Member than the Little Eagle lithofacies. Although the majority of bivalves are also found in the lower Little Eagle this is the first horizon in the Fox Hills sequence in which *P. linguaeformis* is a common element of the fauna; in addition there are some marked differences in the ammonoid assemblages.

The juvenile ammonoids are mostly the young of *D. abyssinus,* and for this reason the concretions are referred to as the *abyssinus* concretions. Mature specimens of *abyssinus* are also relatively abundant. These assemblages are restricted to the interval between the *Nucula* layer and the upper jarositic silt layer only in the Grand River area. Along the Moreau River fossiliferous concretions are exceedingly rare in this interval and the *abyssinus* assemblage occurs in scattered small, chocolate-brown concretions above the upper jarositic silt at several horizons in the lower Timber Lake Member. Rather than being distributed in a well-defined zone that is continuous across changes in type of sediment, the *abyssinus* concretions appear to be restricted to dominantly clayey sand beds in the upper part of the transition into sand and transgress upward in the section across key beds from north to south. In the Moreau Valley area, where the transition to sand is higher in the section, relative to the upper jarosite, than it is along the Grand, the *abyssinus* concretions occur a few feet below the *Sphenodiscus* concretion layer and appear to be coextensive with it. Traced westward along the Moreau, *abyssinus* concretions are found still higher in the section as the lower Timber Lake Member grades laterally into the Irish Creek lithofacies.

DISTRIBUTIONAL FEATURES

The more conspicuous lateral variations in the character and faunal content of the principal assemblage zones and concretion layers in the Little Eagle lithofacies, shown in Figs. 16 through 19, are necessarily generalized because of the limitations imposed by relatively poor exposures. The control along the north and south sides of the Grand River as far east as Little Eagle village, and along the north side of the Moreau is good. Control south of the Moreau and east of South Dakota Highway 63 is not good but is adequate to demonstrate that the assemblages disappear southward as well as westward, for none have yet been found along the Fox Hills bluffs on the south side of the Cheyenne-Moreau divide.

The principal assemblage zones of the Little Eagle lithofacies—the Lower *nicolleti*, *Limopsis-Gervillia*, and *Protocardia-Oxytoma*—almost coincide in their western termination in the Moreau and Grand Valleys and together define a sharp western limit of extent for these rich faunas. To the east, in the easternmost part of the Moreau-Grand divide, around the town of Trail City, fossiliferous concretions become rare or absent in the lower Trail City assemblage zones. In this one area, exposures are too few to determine whether or not this is simply a local condition or whether it indicates an equally sharp eastern limit of distribution for the highly fossiliferous layers. However, the fact that none of the assemblage zones have been found on the east end and along the south bluffs of the Cheyenne-Moreau divide indicates that the highly fossiliferous area does indeed have southwestward-converging eastern and western limits and terminates under the Cheyenne-Moreau divide.

It is the great number of fossils present in the different layers of the assemblages that makes their geographic restriction so obvious. The nature of the outcrop prevents quantitative studies of single fossiliferous layers over any appreciable areas but some estimates can be made of the order of magnitude of these molluscan populations. In the more fossiliferous area of the Lower *nicolleti* Assemblage Zone, experience indicates that a very conservative estimate of the frequency of occurrence of concretions with abundant *S. (H.) nicolleti* is one in every 2,500 square feet. An individual concretion may yield from 10 to 45 individual ammonoids; based on a count of 1,800 specimens, 14 out of every 15 of these will be *S. (H.) nicolleti*. In one 20-square-mile area along the Grand River outcrop, evidence indicates that there are at least 3 or 4 fossiliferous concretions in every 2,500 square feet; even counting only 1 concretion and taking an average of 20 specimens to the concretion there are more than 20 million specimens of *S. (H.) nicolleti* in the 20-square-mile area. For the entire area in which the rich assemblages of *S. (H.) nicolleti* are found the number must be in the hundred millions.

At one locality north of little Eagle a layer of *Gervillia* concretions was exposed over about 300 square feet on the surface of a small bench. Seventeen fossiliferous concretions found in this area contained an average of about 100 specimens each, counting two valves to a specimen: over one square mile this same density would amount to over a hundred million specimens. Unfortunately exposures are not good enough to judge the degree of patchiness of distribution within any assemblage zone, but no matter how conservative one is in manipulating the figures available, the profusion of specimens in the fossil accumulations is truly impressive.

The coincidence in area of the three principal assemblage zones and the occurence of the fourth within it (Fig. 19) point to a recurrence of conditions favoring the accumulation of these great quantities of molluscs within the same general area during an appreciable length of time. As the distribution maps show, the elongate area of the fossil accumulations trends north-northeast. Other features of the Little Eagle lithofacies that conform to this trend are the area of maximum thickness in the interval between the *Protocardia-Oxytoma* concretion layer and the barren A concretion layer, and the distribution of the sand body in its lower part (Fig. 19). The initial appearance of the Timber Lake sand body in the type area, at a horizon somewhat below that of the upper jarosite zone, is in the northeastern part of the area, indicating growth southwestward along the same trend (Fig. 20).

IRISH CREEK LITHOFACIES

The lateral gradation westward of most of the Lower Fox Hills from the Trail City-Timber Lake succession in the eastern part of the type area into the Irish Creek lithofacies is most clearly displayed in the valley of the Moreau River. Here the gradation is complete before the base of the Fox Hills passes beneath the surface just east of the mouth of Irish Creek. The lithofacies was named for exposures near the confluence of this creek and the Moreau, and the type section, described in Section 4, p. 86, is on the south-facing river bluffs just west of the confluence. Additional reference sections illustrating different aspects of the Irish Creek are described in Sections 5 to 7, p. 88 to 90.

The Irish Creek lithofacies, like the Little Eagle lithofacies, consists largely of clay and silt with some fine-grained sand; it weathers to much the same light-gray or silver-gray color on the less precipitous outcrops (Pl. 4, fig. A). Two principal characteristics serve to distinguish it from the Little Eagle: the absence of the abundant fossiliferous concretions of the assemblage zones, and the thin interbedding of the clay and silt throughout much of its lower part, which is the Little Eagle lateral equivalent. The upper part of the Irish Creek, which is equivalent to part of the Timber Lake Member, contains fine-grained sand as well as silt. Here thin interbedding of the clay fraction with the silt or sand is not generally as evident although it occurs locally. Here also, mixing by organisms appears to account for the lack of stratification and swirl structure and filled borings are both common. It is only in this upper mixed portion of the Irish Creek lithofacies that fossils are at all common; these occur chiefly as an extension of the concretion layers of the *Cucullaea* Assemblage Zone of the Timber Lake Member in the Moreau Valley and its lateral variant in the Grand Valley.

The Irish Creek lithofacies is exposed in three separate parts of the type area of the Fox Hills Formation: in the northwest part along the Grand River and its tributaries, on the west side in the Moreau River valley, and on the south along the south side and east end of the Cheyenne-Moreau divide. Lateral changes make it difficult to relate these widely separated areas of outcrop with precision, each area having certain local characteristics not shared with the others. For this reason the Irish Creek stratigraphy is discussed separately for each area. The sequence of key beds in the Lower Fox Hills (Fig. 7) is critical in relating the areas of Irish Creek outcrop to one another and to the Little Eagle lithofacies and Timber Lake Member.

MOREAU RIVER AREA

Lateral gradation from the Little Eagle to the Irish Creek lithofacies starts approximately where S. Dak. Highway 63 crosses the Moreau for it is here that fossiliferous concretions in the Little Eagle assemblage zones begin to become rare. Within 6 miles west of this area the fossil assemblages disappear and concretions are fewer. Some of the individual concretion layers, although largely barren, can be identified by their lithology and relative stratigraphic position. In addition, thin-bedding begins to appear and with it some thin layers of bentonite.

Farther west, in the river breaks east of the crossing of S. Dak Highway 65, the Irish Creek lithofacies is a very sparsely fossiliferous sequence of thinly interbedded silt and silty shale with a few layers of concretions and one or two thin bentonites. In its upper part the thin-bedding generally disappears and it becomes more sandy. The sandy clay and clayey sand extend above the persistent upper jarositic layer to the *Sphenodiscus* concretion layer, indicating a very slight rise in section of the base of the sand of the Timber Lake Member in the 12 or 13 miles that separate the crossings of Highways 63 and 65. Of the lower concretion layers, those at the horizon of the *Protocardia-Oxytoma* Assemblage Zone, though few and rarely fossiliferous, remain calcitic. The small concretions at the Upper *nicolleti* level are common and *Actinosepia* is locally present. The other concretion layers of the lower Trail City have lost their lithologic identity through gradation laterally into hard, purplish-red weathering, sideritic concretions with light brownish-gray, slightly calcitic cores. Higher concretion layers equivalent to upper Little Eagle and lower Timber Lake layers remain calcitic.

The greatest change in the Irish Creek lithofacies takes place abruptly westward within the four miles between S. Dak. Highway 65 and Irish Creek. The Timber Lake Member, which is 70 feet thick a half mile east of Highway 65, completely disappears between it and Irish Creek; in fact, within 2 miles west of the Highway, in the vicinity of St. Patrick's Butte, all that remains of the Timber Lake Member is 15 feet of clayey sand, containing the *Cymbophora-Tellina* Assemblage Zone at the top. The Irish Creek lithofacies has replaced some 50 feet of Timber Lake sand within about 2 miles across the depositional strike. At this point the 15 feet of clayey Timber Lake sand separates the Irish Creek lithofacies from the Iron Lightning Member of the Fox Hills, which, concomitantly, has dropped stratigraphically westward, replacing some of the upper part of the Timber Lake sand body. The Iron Lightning Member continues to drop in the section westward within the next three miles, completely eliminating the Timber Lake and replacing about 25 or 30 feet of the Irish Creek lithofacies, down to and probably into the persistent, locally thick, *Cucullaea* Assemblage Zone.

The contact between the Irish Creek lithofacies and Iron Lightning Member stabilizes at this stratigraphic position throughout the remaining outcrops of this part of the section in Ziebach County. Along the south side of the Cheyenne-Moreau divide it continues to hold this position as far as it has been traced (5 miles) beyond the type area into Meade County. The fossils in the *Cucullaea* Assemblage Zone are amazingly persistent but the zone thins westward and the concretions become sparse. At many places along the western edge of the area only a few scattered concretions with the fauna of the *Cucullaea* Assemblage Zone can be found so that it does appear to be dying out westward. At some places a bentonite or bentonitic shale, the "D" bentonite, occurs just above these scattered *Cucullaea* assemblages and is locally a useful subsidiary key bed where the fossiliferous concretions are few.

The principal features of the transition to the Irish Creek lithofacies along the Moreau River are shown in the correlation chart, Fig. 25. The stratigraphic interval that the Irish Creek lithofacies comes to occupy—from the base of the Fox Hills into or just above the *Cucullaea* Assemblage Zone—increases in thickness westward. At localities along the Moreau around State Highway 63, where the Little Eagle and Timber Lake beds make up the Lower Fox Hills, the interval to the base of the *Cucullaea* Assemblage Zone is about 125 feet. Between State Highway 65 bridge and Irish Creek, where the base of the Fox Hills goes into the subsurface, the same interval is at least 160 feet thick; here, of course, it is all in the Irish Creek lithofacies. This westward thickening is not uniform throughout the Irish Creek lithofacies. In the sections measured along the Moreau that part of the interval between the projected horizon of the *Protocardia-Oxytoma* Assemblage Zone and the upper jarositic zone does not show any consistent thickening or thinning but remains nearly uniform in thickness. The remaining segments of the Irish Creek above and below it show a combined, gradual westward thickening about 3 feet per mile. West of Irish Creek, where the base of the Fox Hills dips beneath the surface, that part of the interval between the persistent upper jarosite and the base of the *Cucullaea* Assemblage Zone alone thickens 18 feet within 6 miles.

The characteristic thin interbedding of shale and silt in the lower part of the Irish Creek is most pronounced west of State Highway 65, chiefly because the northwest dip has brought the beds into the steeper cuts of the inner valley where they weather to rough platey faces rather than rounded clayey bluffs. At State Highway 65 the thin-bedded phase extends at least as high as the horizon of the *Protocardia-Oxytoma* Assemblage Zone and within 10 miles west of the highway it extends to the upper jarosite, the approximate top of the Little Eagle equivalent. How far it rises above this horizon westward is not known. All of the westernmost sections that show the upper part of the Irish Creek consist chiefly of clayey silt and sand with little or no bedding structure in and just below the *Cucullaea* Assemblage Zone; at least throughout the area studied the thin-bedded phase does not come to occupy the entire Irish Creek interval, but only about its lower two thirds.

Key Bed and Fossil Distribution: Changes in the pattern of key beds in the transition to the Irish Creek lithofacies overlap sufficiently to afford adequate stratigraphic control, although not as good as that in the Little Eagle lithofacies. The typical thick-jacketed, silty, light-gray concretions of the Lower *nicolleti* Assemblage Zone, though barren, can be traced westward to sec. 31, T. 15 N., R. 23 E.—about halfway between the Moreau River crossings of highways 63 and 65. From here westward, several concretion layers are common at this horizon but the concretions change to rusty-brown weathering, locally somewhat sideritic, limestone with yellow calcite filling the cracks. About where the change in concretion type takes place a thin bentonite bed, the A bentonite, appears within 4 or 5 feet above the uppermost concretion layer. Commonly only a row of splintery-white, cone-in-cone concretions in a dark-gray bed of bentonitic shale marks this key bed, but from place to place there is a layer of bentonite locally as much as 4 inches thick.

Concretions, generally barren, extend westward from both the *Limopsis-Gervillia* and *Protocardia-Oxytoma* assemblage zones of the Little Eagle lithofacies into the Irish Creek lithofacies at least as far west as Highway 65. Beyond this point they do not appear to persist, or if they do they cannot easily be singled out from the smaller and flatter, unfossiliferous, calcareous to sideritic concretions that are locally a common feature of the lower part of the Irish Creek. The jarositic zone that lies above the *Protocardia-Oxytoma* horizon is more persistent or at least more easily recognized than the concretions beneath it. It can be traced westward until it dips below the surface in the vicinity of Thunder Butte village.

In the beds equivalent to the upper part of the Little Eagle lithofacies both the *Nucula* concretion layer and the generally persistent barren A layer below it extend into the Irish Creek lithofacies. It becomes more difficult to identify these layers westward because of the increasing irregularity of their distribution and the usual local appearance of non-persistent, scattered concretions or local concretion layers at adjacent horizons. The *Nucula* layer appears to be the more persistent, at least its sparse fauna of protobranchs can be found well into the Irish Creek lithofacies. Below this layer and, as far as can be determined, between it and the barren A layer below it, a second bentonite, the B bentonite, appears. The B bentonite begins to be intermittently present along the Moreau little more than a mile west of Highway 63 where its relationship to key horizons in the Little Eagle lithofacies can be seen in the bluffs southwest of the village of Green Grass. It is one of the more persistent key beds in the Irish Creek lithofacies west of Highway 65.

The upper jarosite of the Little Eagle lithofacies, which is used locally to mark the boundary with the Timber Lake member, is a persistent key bed in the Irish Creek lithofacies. At some places it is a group of 2 or 3 jarositic silt layers, at others a highly glauconitic clay or silt. In the beds equivalent to the Timber Lake Member that lie above the upper jarosite three key beds are useful in the Irish Creek lithofacies: 1) the C bentonite, which appears between the upper jarosite and the *Sphenodiscus* concretion layer, 2) the *Sphenodiscus* concretion layer, and 3) the *Cucullaea* Assemblage Zone. The *Sphenodiscus* layer is only sparingly fossiliferous but the concretions are fairly persistent. The C bentonite begins to show up in the sequence about where Highway 65 crosses the Moreau. This set of three key beds is most useful west of the mouth of Irish Creek where the Timber Lake Member is no longer present.

Fossil distribution in the Irish Creek lithofacies is not so strictly confined to concretion layers as in the Little Eagle lithofacies. One outstanding feature is the occurrence of single large specimens of *Sphenodiscus,* usually loose in the shale and silt layers and with concretionary limestone only as fillings in their body chambers; less commonly, *Discoscaphites nebrascensis* is found in this same manner. Large individuals of *Sphenodiscus* do not occur in the equivalent Little Eagle beds to the east, where they are confined to the overlying Timber Lake Member, but they are found here and there throughout the Irish Creek lithofacies. Protobranch bivalves also occur scattered throughout the Irish Creek, in and out of concretions.

Between S. Dak. Highways 63 and 65 the dominantly barren concretion zone at the horizon of the Lower *nicolleti* Assemblage Zone contains rare patches of *Lucina*, both in concretions and surrounding sediment. At the horizon of the Upper *nicolleti* Assemblage Zone the scattered small concretions characteristic of its western phase in the Little Eagle lithofacies persist at least as far west as the Moreau bluffs south of St. Patrick's Butte and have yielded *Actinosepia* there.

That part of the Irish Creek lithofacies equivalent to the Timber Lake Member is particularly fossiliferous in the mile or two of outcrop just west of the abrupt termination of the Timber Lake sand. Here the *Cucullaea* Assemblage Zone is fossiliferous through an interval of about 20 feet, but thins markedly westward. The fauna is dominated by masses of small *Cucullaea;* large specimens, so characteristic of the zone in the Timber Lake sand, are rare. Also common are *Protocardia subquadrata, Pteria linguaeformis,* protobranch pelecypods, several species of scaphites, and *Sphenodiscus.* Perhaps more conspicuous here than in the equivalent Timber Lake concretions is the abundance of wood, charcoal, leaves and nut-like bodies.

GRAND RIVER AREA

The westward transition to the Irish Creek lithofacies along the Grand River begins about 3 or 4 miles west of the longitude of Bullhead. In the segment of the river between Bullhead and the mouth of Firesteel Creek, where most of the change to the Irish Creek lithofacies takes place, the base of the Fox Hills locally rises above and descends below river level. In part this is because the valley wanders at various directions across the regional dip, but small-scale normal faulting and the slumps along the inner valley walls also contribute. This situation coupled with a scarcity of good key beds in the Irish Creek lithofacies makes it difficult to trace the gradation in any detail. The major aspects of the gradation are shown diagrammatically on the correlation chart, Fig. 26.

In the part of the Irish Creek sequence equivalent to the Little Eagle lithofacies the basal jarosite and upper jarosite are both persistent, the latter being the more useful key bed. Of the assemblage zones the *Limopsis-Gervillia* concretions are at least locally fossiliferous as far as 10 miles west-southwest of Bullhead whereas the other assemblage zones are unidentifiable within 3 or 4 miles from Bullhead. Neither the concretions carrying the *Limopsis-Gervillia* association nor the basal jarosite are very helpful key beds because they crop out, near river level, at only a few places. Concretions at the horizon of the *Protocardia-Oxytoma* Assemblage Zone are sparingly fossiliferous along the river about 3 miles southwest of Bullhead (Loc. 253), but beyond this the horizon is marked only by a foot or two of unfossiliferous, jarositic, silty shale that is generally glauconitic and locally contains either small brown-weathering concretions of glauconitic sandstone or small, hard, siliceous concretions with concentric structure.

In beds equivalent to the upper Little Eagle, the upper jarosite is remarkably persistent westward and can be found as far as the bridge of S. Dak. Highway 65, beyond which it apparently dips beneath river level. A persistent zone of brown-weathering, brittle, limestone concretions with yellow calcite as fracture filling lies from 13 to 23 feet below the upper jarosite; it locally contains a few scattered nuculid pelecypods and is a continuation of the *Nucula* concretion layer of the Little Eagle lithofacies. Between this layer and the upper jarosite, fossils locally occur in small concretions and in the sediment. The sparse fauna includes *Pteria linguaeformis, Discoscaphites abyssinus, Sphenodiscus* and protobranchs; it is most likely an impoverished representative of the *abyssinus* association that occurs to the east at the same horizon in the Little Eagle lithofacies.

Nowhere in the segment of the Grand River valley between Bullhead and the Highway 65 bridge, where the transition to the Irish Creek lithofacies takes place, do the beds of the Little Eagle equivalent exhibit the thin-bedded structure that they do along the Moreau. This may be because the transition, exposed for a much shorter distance along the Grand River than along the Moreau, is not complete. Nevertheless the western most outcrops of this part of the Irish Creek, just east of the mouth of Firesteel Creek, show little interbedding of silt and shale and consist predominantly of dark gray, silty to sandy clay that weathers to a blocky or crumbly, light bluish-gray surface. Consistent with the scarcity of bedding, neither the A nor B bentonites have been found in the Irish Creek lithofacies of the Grand River area.

The upper part of the Irish Creek lithofacies, above the upper jarosite, consists of about 60 to 75 feet of dark-gray, clayey, silt and sand. Approximately the lower half is irregularly thin-bedded silty clay or silt, and the upper half is dominantly mixed clayey sand with minor thin bedding. The lower thin-bedded portion is approximately the lateral equivalent of the Rock Creek lithofacies of the Timber Lake Member to the east, and the upper massive portion grades eastward chiefly into the fossiliferous massive sands that overlie this Timber Lake thin-bedded phase. The transition of both parts, from Timber Lake to Irish Creek, takes place rather abruptly in the area between the mouth of White Shirt Creek and Firesteel Creek. Throughout the Irish Creek to the west a fairly persistent layer of concretions lies at the change from the thin-bedded to the massive sediment. Relative to the key bed sequence for the Lower Fox Hills these fall at the horizon of the *Sphenodiscus* concretion layer of the Moreau River area (Sec. 7, p. 90).

In the massive sandy clay 20 to 30 feet above these concretions *Pteria linguaeformis, Cymbophora, Ostrea pellucida* Meek and Hayden, *Protocardia*, protobranchs and *Discoscaphites* locally occur in scattered, punky, sandy concretions with much plant debris. At a few places a thin bentonite bed is preserved at the top of this intermittent zone of fossil accumulation. The fossil assemblage is similar to that found in the lower massive sand of the Timber Lake Member to the east around Bullhead and can be traced directly into it by matching stratigraphic sections along the river. Section 7 of the reference sections, p. 90, from bluffs along White Shirt Creek near its mouth, shows the fossiliferous zone partly in a yellowish-weathering massive sand and partly in dark-gray clayey sand of the intergrading Irish Creek beneath. Here the bentonite is about 12 feet above the base of the 25-foot massive sand and all the fossiliferous concretions occur in that part of the sand below the bentonite. At other localities within a few miles southwest of White Shirt Creek along the river, and also to the west along Hump Creek, the bentonite is a short distance below the top of the Irish Creek facies and just above the highest fossiliferous concretions. The fauna has also been traced 10 miles due south of the Grand River along Firesteel Creek.

This fossiliferous zone and its overlying bentonite correlate respectively with the *Cucullaea* Assemblage Zone and the D bentonite of the Moreau River sequence. The Irish Creek lithofacies along the Grand River has a different association of fossils, lacking *Cucullaea* and having the less varied fauna dominated by *Pteria, Cymbophora* and other genera already noted. The differences between these Moreau and Grand River-Irish Creek faunas are identical to faunal differences that occur between the two valleys to the east in the equivalent lowermost massive sand of the Timber Lake Member, and they occur at the same relative position in the sequence. The D bentonite is more common in the Irish Creek lithofacies of the Moreau Valley but in both areas the faunas lie directly beneath it and do not extend above it. The thickness of the sequence between the upper key jarosite and the D bentonite (or the top of the interval with fossils) varies from 60 to 85 feet in the Grand Valley and from 50 to 65 feet along the Moreau, a difference similar to that between equivalent facies of the lower part of the Timber Lake Member in the two valleys.

The outcrop of the Irish Creek lithofacies on the Grand River goes beneath the surface, about 7 miles due west of Highway 65 bridge, just west of the termination of the Timber Lake Member. Throughout this distance only the uppermost part of the Irish Creek is intermittently exposed so that its relationship to the overlying Timber Lake cannot be determined as accurately as was possible in the area of the Timber Lake pinch-out along the Moreau. As in that area, however, the change to Irish Creek from Timber Lake westward is abrupt, at least the lower 40 feet of the latter passing into Irish Creek between the highway bridge and abandoned village of Black Horse 2.5 miles to the west. The concomitant drop in section of the Iron Lightning Member seems less pronounced but the stratigraphic control is insufficient to estimate it. Together the two facies changes reduce the Timber Lake sand from about 65 feet thick in the vicinity of the bridge to 18 feet at Black Horse. Fossils in the upper 10 or 15 feet of the Irish Creek at Black Horse include *Cymbophora, Pteria, Tancredia* and *Protocardia*. This fauna is estimated to lie about 30 feet higher in the section than that equivalent to the *Cucullaea* Assemblage Zone and is probably equivalent to the *Cymbophora-Tellina* Assemblage Zone of the Moreau River area.

A little over 4 miles west-southwest of Black Horse about 12 feet of dark-gray clayey sand lie between river level and the base of the Iron Lightning Member and contain *Tancredia, Cymbophora* and *Ophiomorpha*. There is nothing distinctive about this fossil assemblage; consequently, where it fits in the Grand River Irish Creek sequence is not known. If the conditions at the west edge of the Timber Lake sand body parallel those found along the Moreau then possibly the Iron Lightning Member has continued to drop in the section west of Black Horse and at this locality lies just above the equivalent of the *Cucullaea* Assemblage Zone. Suggestive of this possibility is the fact that the Iron Lightning Member increases at least 40 feet in thickness between exposures along S. Dak. Highway 65 and the locality in question. But no strong supportive evidence, such as the presence of a bentonite above the fauna, was found and its stratigraphic position remains in question.

In summary, the pattern of change to the Irish Creek facies of both the Little Eagle lithofacies and the lower part of the Timber Lake Member appears to be much the same along the Grand River as it was along the Moreau. Less thin-bedding is evident along the Grand River where it is largely confined to the interval between the upper jarosite and a concretion layer apparently equivalent to the *Sphenodiscus* layer. Only the D bentonite is present in the Irish Creek sequence on the Grand River but both the *Cucullaea* and *Cymbophora-Tellina* Assemblage Zones appear to be represented by associations consisting chiefly of *Cymbophora, Pteria* and *Protocardia*. Because most of the Irish Creek lithofacies dips below the Grand River level before the west edge of the Timber Lake sand body is reached, it is not possible to see whether it thickens westward as it does along the Moreau. Evidence of westward thickening of the Irish Creek where overlain by Timber Lake sand is equivocal. Total thickness of the Irish Creek lithofacies in the vicinity of the mouth of White Shirt Creek is estimated to be about 150 feet; this is based on the projection of the lower part of the section from localities 3 miles to the east and may be somewhat low.

EASTERN AND SOUTHERN PHASE

A glance at the distribution of the Little Eagle lithofacies of the Trail City Member (Fig. 16) shows that there are outcrops of equivalent beds that extend beyond it to the east on the Moreau-Grand divide, and to the east and south on the Cheyenne-Moreau divide. In the latter area these beds are continuous westward with the lower part of the Irish Creek lithofacies. All of these generally unfossiliferous beds in areas east and south of typical Little Eagle have characteristics in common with the Irish Creek lithofacies of the western part of the type area and they are here considered a part of that lithofacies.

One conspicuous difference in the sequence of beds in the eastern and southern Irish Creek lithofacies is that it is equivalent chiefly to the Little Eagle for it is everywhere overlain by the Timber Lake Member. The eastern Irish Creek is delimited primarily by the change from the abundant fossil accumulations of the Little Eagle and secondarily by lithologic change in the concretions and the appearance of a few bentonites. The thin-bedding of silt and clay so characteristic of the Irish Creek in the west is not as conspicuous on the east and south, probably because of the poorer outcrop. In the few good exposures a large part of the sequence shows thin-bedding and lamination of silt and clay on fresh surfaces.

The southern and eastern Irish Creek outcrops are relatively few, small, and widely separated. In both areas the concretion layers can be related directly with concretion layers and assemblage zones in adjacent Little Eagle beds but fossils are very rarely present. In general, the eastern Irish Creek lithofacies is from 5 to more than 20 feet thinner than its Little Eagle equivalent to the west, and a pronounced eastward or southward thinning of the sequence is indicated in the few measured sections.

What little detail is available on the nature of the eastern Irish Creek sequence is shown in Fig. 24. The contact with the underlying Elk Butte Member of the Pierre Shale locally becomes difficult to define as the basal jarosite becomes more discontinuous eastward and the increase in silt content less abrupt. The horizon of the Lower *nicolleti* Assemblage Zone generally lacks the typical, silt-jacketed, light-gray concretions and has instead scattered small, rather nondescript, baseball-sized, jacketed or nonjacketed concretions. The lowermost persistent key bed consists of the highly ferruginous, rusty-brown to purplish-red concretions that mark the horizon of the *Limopsis-Gervillia* Assemblage Zone. At some places cone-in-cone masses indicate the presence of bentonite; few of these are tied in with measured sections but their general stratigraphic position, too far above the base to be the A bed, yet below the Timber Lake contact, suggest that most are the B bentonite.

In the discussion of the Timber Lake Member the increase in clay and the changes in thickness and fauna in the sandstone along the south and east ends of the

Cheyenne-Moreau divide are described. These most likely indicate a transition into Irish Creek lithofacies but the change is not far enough along to make it practical to separate off a portion of the Timber Lake as Irish Creek; it is still predominately yellowish-weathering sand, although very clayey sand, to the east end of the divide.

REFERENCE SECTIONS

Seven sections are given below to illustrate various aspects of the Trail City Member; the map locations are shown on Fig. 15. Section 1 is principal reference for the member, sections 3 and 4 the types, respectively of the Little Eagle and Irish Creek lithofacies. Sections 1 to 3 are in the Little Eagle lithofacies; sections 4 to 7 in the Irish Creek lithofacies.

SECTION 1

PRINCIPAL REFERENCE SECTION—TRAIL CITY MEMBER

Composite section of the Trail City Member taken chiefly from exposures east of South Dakota Highway 63, in E ½, SW ¼, sec. 32, T. 15 N., R. 24 E., U.S.G.S. Parade NW quadrangle, Dewey County. Top of section (unit 20) is caprock of small butte in NE ¼, NE ¼, SW ¼, sec. 32, and most of section was derived from its SW flank (Loc. 196) and along spur that extends due S of its base (Loc. 199). Supplementary information, particularly on thickness of intervals, was derived from exposures just west of Highway 63 along the west edge of section 32 (Loc. 30)

	Thickness (feet)

Lower Fox Hills Formation (in part):
Timber Lake Member (in part):

20. Concretion layer in sand; sandy limestone concretions up to 2 ft. diam., weather reddish brown; some jacketed in calcareous silt and sand; predominantly barren, some with single *Sphenodiscus* (*Sphenodiscus* concretion layer) 2

19. Sand, very fine-grained, variably clayey, chiefly massive, weathers yellowish gray; locally upper 8 feet contains scattered, hard, highly fossiliferous concretions up to 1 ft. diam. that weather deep reddish brown, some characterized by abundant juvenile scaphites (*abyssinus* concretions) 20

18. Sand, very fine-grained, clayey, becoming silty in lower half, weathers yellowish gray; at top scattered small gray-weathering limestone concretions with vuggy purple-red interiors 12

Trail City Member.

17. Clay, silty and sandy, mixed with clayey sand, weathers gray 4

16. Sand, very fine-grained, glauconitic, weathers orange brown; punky, sandy, calcareous concretions in lower part; locally fossiliferous. (=upper jarositic silt layer) 3

15. Clayey silt and sand, weather light gray with some yellowish-gray stain 7

14. Concretion layer; brittle, ovoid, limestone, weathers rusty brown, 1.5 to 3 ft. in long diam., cracks coated with yellow calcite; some with silty jackets; chiefly barren (*Nucula* concretion layer) 2

13. Silt, clayey to shaly, gray, with finely silty fissile shale locally in lower 5 ft.; upper 8 ft. becomes increasingly sandy........................... 16

12. Concretion layer; brittle, flat-ovoid, limestone, up to 1.5 ft. diam., weathers gray to brownish gray, some yellow calcite in cracks (barren A concretion layer) 1

11. Shale, silty, weathers gray to brownish gray, checked surface; becoming siltier in upper part, in sharp contact with unit below 1

10. Silt, clayey, mixed, some local shaly lenses; scattered blebs and patches jarositic silt throughout and concretions jarosite at top and base; weathers light brownish gray to yellowish gray; a few scattered silty limestone concretions with thick, punky, some cone-in-cone, jackets (medial jarositic silt zone) 9

9. Clayey silt, mixed, weathers light gray 5

8. Sand with limestone concretions; very fine-grained, rusty-weathering, slightly glauconitic sand with scattered concretions about 1 ft. diam. in upper part; some weather red brown with gray silty jackets, some lack jackets, weather cream-colored. Locally highly fossiliferous (*Protocardia-Oxytoma* Assemblage Zone) 2

7. Clayey silt, locally grades laterally to thinly interbedded silt and silty clay; contains scattered small, 4 to 10 inch diam., round to ovoid concretions, few fossils, chiefly *Discoscaphites*, sepioid remains, and locally, calcified trunks of *Palmoxylon* (*Actinosepia* concretions) 9

6. Clayey silt, mixed, somewhat shaly in upper part, weathers light gray 10

5. Concretion layer, limestone concretions, about 1 ft. diam., weathering rusty brown, locally with accumulations dominated by *Limopsis* (*Limopsis-Gervillia* Assemblage Zone) 1

4. Clayey silt, mixed, weathers light gray 9

3. Clayey silt, mixed, weathers light gray, with scattered small (up to 6 inch) concretions, round to flat-ovoid, weather light brownish gray, gray and cream-colored; a few with gray silt jackets, rare single fossils chiefly *Scaphites* (*Hoploscaphites*) *nicolleti* 5.5

2. Clayey silt, mixed, weathers light gray, thin bands of limonite-jarosite stain at base and 6 ft. above base; in upper 1.5 ft. scattered limestone concretions, weather light gray to pinkish gray, some with punky silt jackets; rare scattered specimens S. (*H.*) *nicolleti* 7.5
                                                                         —— —
    Total thickness Trail City Member .... 98 feet

Pierre Shale:
Elk Butte Member:

1. Shale, silty, weathers gray to light gray, becoming increasingly silty in upper 8 ft. .... 10 to ?

The Trail City in this section, though in the Little Eagle lithofacies, is beginning to change to Irish Creek. It is thicker than average and its principal assemblage zones of fossils are atypical in a number of characteristics. The Lower *nicolleti* zone, which here includes units 2 and 3, is thick but has relatively few of its characteristic round to ovoid, gray, limestone concretions with the thick, punky, silt jackets, and only one of the highly fossiliferous concretions have been found here or within 6 to 8 miles east of this section. Instead the zone is characterized by the small concretions a few of which include single specimens of ammonites (see Pl. 2, fig. C). The *Limopsis-Gervillia* Assemblage Zone is reduced here to a single layer of widely spaced concretions, typically rusty on weathered outcrops, few of which contain masses of *Limopsis*.

The *Protocardia-Oxytoma* Assemblage Zone is the most fossiliferous in the vicinity of the section, but it too is relatively sparsely fossiliferous, and accumulations of *Protocardia* and *Oxytoma* are fewer. They are in part replaced locally by an assemblage in which

Lucina is the most characteristic bivalve, and also by assemblages consisting dominantly of ammonoids. On the other hand, in the underlying interval (unit 7), the small concretions containing isolated scaphites and rare *Actinosepia* have been more productive in this area than at most localities.

Perhaps the greatest advantage of this area along S. Dak. Highway 63 as principal reference area for the Trail City Member is its fine display of lateral variation in the member from exposure to exposure. The largest single exposure (Loc. 30) is in the forked head of a deep gully around which the road bends to the east, in the NE ¼, SW ¼, NW ¼, sec. 32, T. 15 N., R. 24 E. (see Pl. 1). Here continuous exposure of all but the upper 15 feet of the member from its base to the Barren A concretion layer, permits detailed study of the interval and its concretions in relatively unweathered condition. The base is obscure here because the change from shale to clayey silt is less abrupt locally than in areas southeast of the road, and the basal jarositic zone is absent. By contrast the medial jarositic zone (Unit 10) is well defined in the upper slopes of the gully and serves to orient one in the section. This prominent band of yellowish-weathering silt was mistaken by Morgan and Petsch (1945, p. 24) for the one marking the base of the Trail City at this locality.

Other exposures in the vicinity are more weathered and in them can be seen minor variations in thickness between key beds, in the lithology of the units and in their weathered aspect. The fossil content of the concretions in the *Protocardia-Oxytoma* and *Limopsis-Gervillia* zones increases to the east and southeast of the road and the beds containing the *Protocardia-Oxytoma* concretions become sandier in this direction.

SECTION 2

Although the area about Section 1 is fairly representative of the less fossiliferous upper Trail City beds, it is definitely not typical of the lower part of the member in the Little Eagle lithofacies, from the *Protocardia-Oxytoma* concretions down. Characteristic, fossiliferous beds of this part of the lithofacies are fairly well exposed in local bluffs for a considerable distance on either side of the road from Timber Lake to Whitehorse, where it crosses the south edge of the Moreau-Grand divide at the east edge of sections 24 and 25, T. 16 N., R. 25 E. East of the road in T. 16 N., R. 26 E., the bluffs extend through sections 30, 19, 20, and 28 in a prominent, arcuate bluff, hereafter referred to as Whitehorse Ridge, that can be used as a reference area for the lower part of the Trail City and Little Eagle lithofacies in the Moreau Valley; it lies approximately 16 miles southwest of the village of Trail City. The following section is typical of exposures in this area.

Composite section from exposures of the lower part of the Trail City Member in the central part of Whitehorse Ridge; chiefly from southwest slope of small butte in W ½, NW ¼, SW ¼, SW ¼, sec. 20, T. 16 N., R. 26 E., U.S.G.S. Whitehorse quadrangle, Dewey County (Loc. 8) but supplemented by data from adjacent exposures in the N ½, SW ¼, sec. 20. (Locs. 6 and 7)

Thickness
(feet)

Lower Fox Hills Formation (in part):
Trail City Member, Little Eagle lithofacies (in part):

8. Concretion layer in sand; large, flat-ovoid, limestone concretionary masses up to 4 ft. diam. and 10 inches thick and smaller ovoid concretions 10 inches, or less, diam., weather red brown and gray, some of latter very fossiliferous (*Protocardia-Oxytoma* Assemblage Zone). Caps local butte and adjacent benches .... 1 to ?

7. Sand, very fine-grained, friable to clayey, weathers light yellowish brown, grades downward to sandy silt in lower half. Between 7 and 10 feet from

top are scattered limestone concretions about 1 ft. diam., weather red brown, a very few contain abundant ammonoids (Upper *nicolleti* Assemblage Zone) 18

6. Clayey silt, mixed, weathers light gray 7

5. Concretion layer in clayey silt; scattered, ovoid, limestone concretions, 1 ft. or less diam., weather rich rusty brown, a few with abundant *Limopsis* (*Limopsis-Gervillia* Assemblage Zone) 1

4. Clayey silt, mixed but locally with unmixed lenses silty gray shale and gray-white silt; silt pods locally jarositic; weathers light gray to light brownish gray 7

3. Clayey silt as in unit above, jarositic blebs throughout but more concentrated in lower 3 feet; several layers large (over 1 ft.) ovoid, gray, limestone concretions with thick punky gray silt jackets; a few in upper layer with abundant *S. (H.) nicolleti* (Lower *nicolleti* Assemblage Zone) 5

Total Trail City Member measured .... 39 feet

Pierre Shale:
Elk Butte Member:

2. Clay, silty, becoming increasingly silty upward; grades into shale below. . 4

1. Shale, silty, dark gray with brown discoloration on fractures, weathers gray; scattered small, ovoid, gray limestone concretions at top .... 10 to ?

Concretions in the Upper *nicolleti* Assemblage Zone are generally barren along Whitehorse Ridge except in very local areas where a few fossiliferous ones are found fairly close together. At some places the concretions of this zone occur within 4 or 5 feet of the concretions of the overlying *Protocardia-Oxytoma* Assemblage Zone and on poor outcrops the float from this latter zone may completely obscure the presence of the Upper *nicolleti* concretions.

The base of the Trail City Member is fairly distinct all along Whitehorse Ridge and the basal jarosite is common, though not everywhere present. As in the reference section above, the basal jarositic zone (lower 3 feet of unit 3) is generally underlain by 4 to 8 feet of beds transitional into typical Elk Butte shale.

SECTION 3

In the Grand River valley exposures of the Little Eagle lithofacies of the Trail City are more abundant than along the Moreau River and at a number of places complete sections can be pieced together from adjacent outcrops. Chosen as type section of the Little Eagle lithofacies is an excellent exposure about 4 miles west of the village of Little Eagle, less than a mile north of the Grand River. Here in a single bluff face the entire lower part of the Trail City Member, from 15 feet or more below its base to 10 feet above the *Protocardia-Oxytoma* concretions, is exposed, and the remainder of the section can be pieced from nearby exposures.

TYPE SECTION, LITTLE EAGLE LITHOFACIES

Southwest-facing bluffs along narrow end of southeast-trending spur in SW ¼, sec. 26, T. 20 N., R. 26 E., U.S.G.S. Little Eagle NW quadrangle, Corson County (Loc. 50). Units 1 through 8 were measured on SW face of bluff at end of spur in center of SW ¼ sec. 26. As this is on upthrown side of small NE-trending fault, an additional 40 feet of section, units 9 through 14, are exposed just NW of this face along the bluff. The three upper units are in a partly grassed and sloughed slope on a higher part of the spur just east of the west line of section 26.

Thickness
(feet)

Lower Fox Hills Formation (in part):
Timber Lake Member (in part):

17. Sand, fine-grained, weathers light yellowish brown; contains lenticular, brown-weathering, concretionary masses and ovoid concretions 1 to 1.5 ft. thick .... 15+

16. Partially obscured; sand and clayey sand, weathers yellowish gray; some gray silty shale layers 15

Trail City Member, Little Eagle lithofacies:

15. Sand, shaly, weathers greenish gray with checked surface; may be bentonitic (questionably the horizon of upper jarositic silt)................. 1

14. Clayey sand and silt, scattered irregular layers and blebs dark gray silty shale; weathers yellowish gray; between 1 and 2.5 feet from top is zone scattered, small (6 inch diam.), round to ovoid, hard, blue-gray limestone concretions, weather rusty brown; sparingly fossiliferous (abyssinus concretions) 10

13. Concretion layer; scattered, large (up to 3.5 ft. long diam.), ovoid limestone concretions weathering yellowish gray to brown, with yellow calcite in cracks; locally unfossiliferous except for plant fragments (Nucula concretion layer) 2

12. Clayey silt, slightly sandy, chiefly mixed but locally silty shale; weathers gray to light gray 10

11. Clayey sand and silt, silty clay, minor amounts of shale, mixed, scattered plant fragments weathers yellowish gray; at top, layer flat-ovoid limestone concretions about 1 ft. long diam., weather gray to light gray (barren A layer) 5

10. Shale, silty to finely sandy with laminae and irregular blebs of silt, scattered plant fragments 2.5

9. Sand, very fine to fine grained, glauconitic, mixed with silt and clay; mottled gray and greenish gray, weathers yellowish gray; upper foot is crumbly, Fe-impregnated rusty brown silty and sandy ledge 5.5

8. Concretion layer in sand; crumbly, rusty-brown to reddish-weathering sandy ledge studded with hard limestone concretions, ovoid to irregular, about 1 ft. diam., weather red brown, generally with tough sandy jackets; fossiliferous (Protocardia-Oxytoma Assemblage Zone) 2

7. Sand, very fine to fine grained, clayey and silty, glauconitic; mixed except for lower 0.5 ft. which is laminated 4

6. Clayey silt and sand, becoming more clayey downward, upper 3.5 ft. crudely thin-bedded, lower 2.5 ft. mixed; jarositic silt blebs 2.5 ft. from base 6

5. Clayey silt, minor fine-grained sand and gray shale, mixed at top and base but irregularly thin-bedded and laminated in interval 2.5 to 7.5 ft. from base. Basal foot contains scattered, rusty-weathering, jacketed limestone concretions, flat-ovoid up to 2.5 ft. long diam., with accumulations of large Gervillia and Limopsis. Upper 6 feet contain scattered, flat-ovoid, gray-weathering limestone concretions with accumulations of small Gervillia (Limopsis-Gervillia Assemblage Zone) 9.5

4. Clayey silt, mixed, weathers light yellowish gray 9

3. Clayey silt, mixed, scattered jarositic silt blebs, weathers light gray. Upper 2 ft. contains scattered gray limestone concretions, flat-ovoid, with thick calcareous siltstone jackets and black gypsiferous inner rinds; weather light gray, some contain abundant S. (H.) nicolleti (Lower nicolleti Assemblage Zone—in part) 1

2. Shale, silty to very silty, mottled gray and dark gray; jarositic silt blebs throughout, common in upper 1.5 ft., and concentrated in upper 0.5 ft. Upper foot contains scattered, flat-ovoid concretions with punky gray silty limestone cores and thick light-gray jackets calcareous silt; locally fossiliferous (Lower *nicolleti* Assemblage Zone in part) 3.5

Total thickness Trail City Member .... 77 feet

Pierre Shale:
Elk Butte Member:

1. Shale, silty, gray, blocky to platey, brownish stain on fracture surface ... 20 to ?

In the partially obscured upper part of this section the peculiar greenish shaly sand of unit 15 was arbitrarily chosen as the top of the Little Eagle lithofacies because it is at the approximate horizon of the upper jarositic silt. The overlying unit 16 is not well enough exposed to determine whether sand dominates the entire interval; part of it may belong in the Little Eagle. The base of the Little Eagle illustrates a fairly uncommon variation for the Trail City Member in that the shaly structure of the underlying Elk Butte persists into the basal jarositic zone (unit 2).

SECTION 4

The Irish Creek lithofacies of the Trail City Member is best exposed in the bluffs and cutbanks along the Moreau River from the vicinity of S. Dak. Highway 65 to the village of Iron Lightning. No single section of it is complete from the contact with the Pierre Shale to that with the Iron Lightning Member. Section 4, which lacks both the top and bottom contacts but is one of the thickest and freshest exposures, is here designated as the type section of the lithofacies. It is supplemented by sections 5 and 6 which illustrate the nature of the lower and upper contacts respectively.

TYPE SECTION, IRISH CREEK LITHOFACIES

South-facing cutbank and bluff north of the Moreau River about 0.8 miles southwest of its confluence with Irish (formerly Worthless) Creek, in the center, sec. 32, T. 15 N., R. 21 E., U.S.G.S. Dupree NE quadrangle, Ziebach County, South Dakota (Loc. 100). Measurements of two exposures just E. of the center of sec. 32 were combined by tracing bentonite layers.

Thickness
(feet)

Gravel-capped terrace

Fox Hills Formation (in part):
Trail City Member, Irish Creek lithofacies (in part):

16. Clay, sandy, gray, some brownish stain (from overlying gravel); at top, persistent layer widely spaced limestone concretions between 1 and 2 feet in diameter, some with siltstone rinds; locally with fossils *Sphenodiscus, Discoscaphites,* a few *Cucullaea* (*Sphenodiscus* concretion layer) .   10

15. Bentonite, with cone-in-cone masses (C bentonite) 1.2

14. Clay, silty to finely sandy, dark-gray; locally some thin interbedding but not conspicuous   16

13. Clay, sandy in lower 3 feet, with scattered jarosite blebs; grading to silty, dark-gray clay, locally laminated, shaly, thin jarosite 6.5 feet from base; at top is 0.3 silty sand heavily impregnated with jarosite and limonite, forms crumbly little ledge (upper jarosite) 23.8

12. Clay, very sandy, dark-gray, weathers gray; with persistent layer sandy limestone concretions at top, weather yellowish brown, very sparsely fossiliferous. (*Nucula* layer) 3

11. Bentonite, with cone-in-cone masses (B bentonite) 0.7

10. Clay, sandy, and clayey sand, weathers light gray; in upper 0.7 feet tough gray clayey sand with clay blebs 7.3

9. Clay, sandy and clayey sand, mixed, with blebs of sand, some jarositic; at 7 feet above base rusty silt with punky concretions; zone of jarositic clayey silt 11 to 12 feet from base 19

8. Clay, sandy, with irregular blebs of sand, grading downward to silty clay; mixed, weathers light brownish gray 9

7. Clay, dark-gray, and light gray silt; thinly, irregularly, interbedded and interlaminated, some local organic mixing of clay and silt; some sand with silt; at 9 feet above base layer flat-ovoid, rusty-weathering limestone concretions, barren; at top persistent layer, rusty-brown-weathering limestone concretions, rare fossils, small protobranch bivalves, *Dentalium*, ammonite fragments (probably *Protocardia-Oxytoma* level) 20

6. Clay, dark-gray, and light gray silt, thinly interbedded; beds locally as much as 0.4 foot thick; scattered burrows; some lensing of beds; at top, layer flat-ovoid, brown- to orange-brown weathering, barren limestone concretions 12

5. Clay and silt as in unit 6 above; at top few scattered limestone concretions with silt, cone-in-cone rinds 7

4. Clay and silt as in unit 6 above; at top, layer thin, orange-brown-weathering, limestone concretions 4

3. Clay, dark-gray, and light gray silt, thinly interbedded and interlaminated, beds irregular, locally lenticular, much cross-lamination; burrows or organically worked patches very rare; layer scattered, flat-ovoid, orange-brown-weathering limestone concretions at top 7.5

2. Clay and silt as in unit 3 above; at top, layer flat, orange-brown-weathering limestone concretions interspersed with scattered, larger, ovoid concretions with siltstone rinds 2.5

1. Clay and silt as in unit 3 above; at top, layer flat, silty limestone concretions with silt rinds weathering brownish gray 3

Total Irish Creek lithofacies measured .... 146 feet

Water level, Moreau River

Correlation of the lower part of this section is not clear, the B bentonite is the lowest horizon that can be identified with certainty. Two and one-half miles east of this section the base of the Irish Creek lithofacies in section 5 is in contact with Elk Butte lithology at a level that can be traced farther east into the jarosite marking the base of the Little Eagle lithofacies. But in section 4 above, neither the jarosite nor the iron-stained silt and A bentonite useful in section 5, are present to indicate this level; instead thin-bedded silt and clay extend downward to water level without change to Elk Butte lithology. Rapid westward thickening of the lower part of the Irish Creek may have carried the level below the surface, or the Irish Creek lithofacies may have begun to replace the Elk Butte Member to the west. Either possibility or a combination of the two would explain the relatively abrupt disappearance of Elk Butte lithology between section 5, where over 50 feet of it are exposed above river level, and section 4, 2.5 miles to the west, where there is none. Structural changes can be eliminated as a factor because of the persistence of the distinctive fossiliferous concretions of the *Cucullaea* Assemblage Zone at the same elevation (approximately the 2200-foot contour) in bluffs around both sections 4 and 5.

SECTION 5

Southwest-facing bluff above meander scar north side Moreau River in SW ¼, SW ¼, NW ¼, sec. 35, T. 15 N., R. 21 E., U.S.G.S. Dupree NW quadrangle, Ziebach County, South Dakota (Loc. 128). Section begins in upper part of slope about 35 feet below base of very large concretions near top of exposure.

Thickness
(feet)

Fox Hills Formation (in part):
Trail City Member, Irish Creek lithofacies (in part):

7. Clay, silty, gray and light-gray silt, thinly interbedded and interlaminated; upper 5 feet some organic mixing, jarosite blebs and scattered small, red-brown, ovoid limestone concretions; at top, persistent layer scattered red-brown-weathering, limestone concretions about 1 foot diameter 25

6. Covered by slope wash 16.5

5. Clay, silty, gray and light-gray silt; irregularly, thinly interbedded; thin beds predominate in lower 3 feet which weathers with deep rusty-brown stain; beds above less regular with pods, layers and irregular blebs of silt in shaly clay; at 6.5 from base several inches bentonitic shale with scattered cone-in-cone concretionary masses (A bentonite) 10.5

4. Clay, silty, with irregular layers and blebs light-gray silt; at top, layer gray limestone concretions up to 1 foot diameter with silty, locally cone-in-cone rinds 2

3. Clay, hard, mottled-brown and greenish-gray with phosphatic nodules .. 0.3

2. Clay, silty, dark-gray, becoming shaly downward; some scattered layers light-gray silt; 3 feet from base is layer calcareous concretions, about 1 foot diameter, weathering light brownish gray 3.7
                                                                    ——
     Total Trail City Member measured .... 58 feet

Pierre Shale (in part):
Elk Butte Member (in part):

1. Shale, dark-gray, very finely silty, blocky, brownish stain on fracture surfaces, weathers to small flakes; scattered red-weathering small sideritic concretions; a persistent layer about 12 feet from top .... 27 to ?

Top of slope wash (about 35 feet above river level).

The change from typical Elk Butte lithology (unit 1) to Irish Creek lithology is completely gradational and choice of the contact arbitrary within a number of feet as the silt appears gradually in the sequence. The phosphatic nodule layer was found only at this locality. Concretions in units 2 through 4 are probably equivalent to the Lower *nicolleti* zone.

SECTION 6

Composite from several sections measured in bluffs south of the Moreau River about 1 mile SE of Thunder Butte village, between the forks of the county roads leading NW to the village and N to the bridge across the Moreau River, in S ¼, NE ¼, sec. 3, T. 14 N., R. 20 E., U.S.G.S. Thunder Butte quadrangle, Ziebach County, South Dakota (Loc. 73).

Thickness
(feet)

Flat on top of bluffs
Fox Hills Formation (in part):
Iron Lightning Member (in part):

10. Shale, silty, tough, gray to brownish-gray and thin interbeds shaly silt,

abundant finely comminuted plant fragments on bedding surfaces; locally weathers to hard checked ("popcorn") crust suggesting it is bentonitic; 3 feet from base orange-brown silt layer with flattened, small, red-brown-weathering, silty, limestone concretions .... 6 to ?

Trail City Member, Irish Creek lithofacies (in part):

9. Sand, light-gray, very fine-grained, thinly interbedded with gray to dark gray shale with carbonized plant fragments; jarosite concentrations at top and base in sand 2.2

8. Shale, black, very finely silty, becoming grayer and more fissile upward ... 6.3

7. Silt, clayey, grading upward to dark-gray silty clay, chiefly mixed but shaly at top with silt partings; small, ovoid, limestone concretions in basal foot sparingly fossiliferous 8

6. Bentonitic clay, greenish-gray, scattered cone-in-cone masses (D bentonite) 0.5

5. Silt, clayey, weathers brownish gray, organically mixed, few local thin-beds or laminae of silt; scattered small limestone concretions, some fossiliferous, and in lower 2 feet persistent zone highly fossiliferous concretions under a foot in diameter; most commonly *Cucullaea* (small), *Protocardia*, *Discoscaphites*, *Pteria*, numerous gastropods, wood with boring bivalves (*Cucullaea* Assemblage Zone) 10

4. Silt, clayey, dark-gray, weathers light gray, gypsiferous in lower 8 feet; above weathers with brownish stain; some thin-bedding of silt and silty clay 17

3. Silt, clayey, dark-gray, weathers light gray, local obscure bedding of silt and silty clay, mostly mixed; at top, persistent layer of large, flat-ovoid, light-brown-weathering calcareous concretions with yellow calcite fracture filling; barren (= *Sphenodiscus* concretion layer) 19.5

2. Bentonite, light gray-green with masses cone-in-cone (C bentonite) ..... 0.5+

1. Silt, gray, clayey, chiefly massive, weathers light gray .... 10 to ?

Total Irish Creek measured .... 74 feet

Slope wash.

SECTION 7

This section illustrates the nature of the Irish Creek lithofacies of the Grand River area; parts of the Irish Creek in this area are also described at the base of sections 11 and 12.

Exposure on east-facing bluff above White Shirt Creek just west of its confluence with the Grand River in center, W ½, NE ¼, sec. 34, T. 21 N., R. 23 E., U.S.G.S. Black Horse NE quadrangle, Corson County, South Dakota (Loc. 254). (From measurements by I. G. Speden and W. A. Cobban)

Thickness
(feet)

Top of bluff

Fox Hills Formation (in part):
Timber Lake Member (in part):

6. Sand, fine- to medium-grained, weathers yellowish orange, thinly interbedded and interlaminated with sandy, gray clay; sand layers up to 2 feet thick; few scattered *Tancredia* .... 5±

5. Sand, fine- to medium-grained, massive, weathers yellowish orange to yellowish gray, forms prominent ledge; sandy clay parting 0.5 foot thick 12 feet from top; bentonitic clay 0.5 foot thick 13 feet from base

(=D bentonite); scattered limestone concretions in lower 10 feet, some fossiliferous, with large red-brown-weathering tabular to irregular, locally fossiliferous concretionary masses at base; *Pteria, Ostrea, Cymbophora, Panopea, Lunatia, Discoscaphites* (= *Cucullaea* Assemblage Zone).   34

Trail City Member, Irish Creek lithofacies (in part):

4. Sand, clayey, silty, dark-gray, and silty shale, weathers gray, massive to faintly laminated; scattered small, tabular sandy concretions in upper 4 feet; few scattered fossils, chiefly *Pteria, Protocardia, Discoscaphites*   ...   30

3. Sand, clayey, dark-gray and silty shale, thinly interbedded and interlaminated; at top, layer brown-weathering limestone concretions with silt rinds. (= *Sphenodiscus* concretion layer)   44

2. Clay, sandy, impregnated with jarosite, some lenses sand. (upper jarosite) 1.5

1. Clay, sandy and silty, dark-gray, massive, scattered concretions brown-weathering limestone in upper 2 feet, some fossiliferous—*Pteria, Discoscaphites abyssinus, Sphenodiscus* (= *abyssinus* concretions); about 12 feet from top layer large, brown-weathering, calcareous concretions with yellow calcite in fractures (= *Nucula* layer) 18.5
    —————
Total Irish Creek measured   ....   94 feet

Creek level.

In this section the change from Timber Lake to Irish Creek lithofacies is fairly sharp. Followed laterally the fossiliferous concretionary masses at the level of the *Cucullaea* Assemblage Zone are in the yellowish sand of the Timber Lake at some places and in the gray clayey sand of the Irish Creek at others. The interval (unit 3) between the upper jarosite and the concretions equivalent to the *Sphenodiscus* layer is the only markedly thin-bedded interval in the Irish Creek lithofacies of the Grand River area.

Timber Lake Member

DISTRIBUTION

The Timber Lake Member, the sand lithofacies of the Lower Fox Hills, forms the surface of the eastern ends of the divides between the Cheyenne, Moreau and Grand Rivers. Here wind action on the poorly consolidated sandstone during periods of drought has excavated numerous, shallow, undrained depressions that dot the relatively flat upland. These depressions are a topographic trade mark of the Timber Lake Member and, together with cultivated fields, form a pattern characteristic of the member on aerial photographs. Outcrops of the sandstone are limited chiefly to gully and valley walls, road cuts, and rare grass-free patches on the sides of buttes and the more precipitous bluffs along the edges of divides. The step-like descent of the topography eastward nearly opposite to the direction of dip, restricts the Timber Lake outcrop on the east side of the area to the relatively flat divides. To the west, where the outcrop narrows toward the rivers and steeper valley walls afford better exposures, the member disappears. These factors drastically limit the possibility of finding good outcrops of more than a part of the Timber Lake Member at any one locality.

Morgan and Petsch (1945, p. 15) named the member for exposures around the town of Timber Lake, county seat of Dewey County on the Moreau-Grand divide. A few outcrops, none showing more than 20 feet of yellowish-orange, weathered sand, occur on the northeast side of town around the large, commonly water-filled, shallow depression from which the town takes its name (in the literature prior to 1912 this depression is called Soda Lake). No adequately exposed section of a significantly large part of the member can be seen within a 10-mile radius of the town although its general lithologic character is evident in numerous road beds and gutters.

No specific exposures were designated as the type either by Morgan and Petsch or by subsequent workers. To date the member has been used as a map unit by the State Geological Survey in eight 15-minute geologic quadrangle maps covering parts of northeastern Dewey and eastern Corson Counties. The original name Timber Lake Sandstone Member (Morgan and Petsch, 1945) was subsequently shortened to Timber Lake Member (Stevenson, 1956, 1957, 1959, 1960a, b; Pettyjohn 1961).

The generally held belief that the Timber Lake Member is continuous throughout the type and adjacent areas of Fox Hills outcrop is a misconception. It disappears completely, by facies change, in the western part of the area. Northeastward from the type area it thickens at the expense of the underlying Trail City Member, also by lateral facies change, and in the Missouri Valley area of North Dakota and adjacent South Dakota it is doubtful whether enough of the underlying Trail City lithology is left to make practical the subdivision of the Lower Fox Hills into two members. Consequently the Timber Lake Member is a useful map unit chiefly in the eastern two thirds of the type area of the Fox Hills where it is commonly over 80 feet thick and encroaches but little on the underlying Trail City Member.

Recognition of the Timber Lake sandstone as a distinctive lithofacies is the principal concern here, regardless of whether it is too thin to map or whether it occupies nearly all of the Lower Fox Hills interval. The informal terms Timber Lake sand, sand body and lithofacies are used frequently in this broader sense in the following discussion: throughout much of the type area these are, of course, synonymous with Timber Lake Member.

CONTACTS

The gradational lower contact of the Timber Lake sandstone descends stratigraphically northeastward. In the northeast corner of the type area, in the bluffs above the Missouri River east of McLaughlin, less than 10 feet of clayey beds separate the Timber Lake and the sandy beds associated with the *Protocardia-Oxytoma* Assemblage Zone of the underlying Trail City Member. Just to the northeast, and probably also to the north, of the type area this lower sand coalesces with the Timber Lake lithofacies, the base of which is at or below the *Protocardia-Oxytoma* level through the contiguous Fox Hills outcrop in the Missouri Valley of North Dakota.

Southwestward from the northeast corner of the type area, in the vicinity of Little Eagle, the base of the Timber Lake sandstone lies approximately at the upper jarosite layer used locally to mark the Timber Lake—Trail City contact. Here the change from the clayey Trail City to Timber Lake sandstone is relatively abrupt. Except for localities along High Bank Creek 10 miles southwest of Little Eagle, and northeast of the crook of Grand River where it turns abruptly north toward Bullhead, this area around and northeast of Little Eagle is the only part of the type area in which sandstone, rather than a transitional clayey sand phase, marks the base of the Timber Lake lithofacies. To the east, south and west dominantly clayey sand marks the transition downward into the clayey Irish Creek and Little Eagle lithofacies of the

Trail City. At most places this lower Timber Lake transition consists largely of mixed sand, clay and silt, but in the vicinity of Bullhead on the Grand River it consists of a distinctive thin-bedded sequence of silty shale and sand. Both kinds of transitional phase are generally between 10 and 50 feet thick. Where they overlie the Little Eagle lithofacies the base of each is usually within 10 feet above or below the upper jarosite and the latter, because of its persistence, is a satisfactory, if arbitrary, base for the Timber Lake Member. West of the limits of the Little Eagle lithofacies the clayey Irish Creek lithofacies appears at a progressively higher position in the section as the Timber Lake transition beds grade laterally into it; here the key upper jarosite bed lies within Irish Creek lithology and the base of the Timber Lake Member is best taken where sand obviously becomes predominant.

The upper contact of the Timber Lake is more sharply defined by lithologic change than is the lower contact but it is much more rarely exposed. Change to the distinctive, brownish-weathering, thinly interbedded sand and shale of the Bullhead lithofacies of the Iron Lightning Member is abrupt and can be pinpointed at most outcrops. In the west half of the type area the contact can be demonstrated to descend stratigraphically southwestward (see Fig. 25).

The series of sections along the Moreau River (Fig. 25) indicates that the upper part of the Timber Lake sand is a lateral facies of the Iron Lightning Member. Throughout the type area the contact itself is sharp, featuring a planed surface and some incorporation of the Timber Lake sand in the basal Bullhead lithofacies, usually in one or two beds from 4 to 12 inches thick. Locally in exposures along the north side of the Moreau River Colgate-type sand is apparently in contact with the Timber Lake sand and appears gradational except for a thin clay-pellet zone. This relationship is discussed in more detail in a later section. Contact of the Timber Lake and Iron Lightning Members indicates an abrupt change in depositional environment, but no aspect of the stratigraphy or fossil distribution suggests an appreciable interval of erosion or nondeposition.

THICKNESS

The relatively few exposures of the Timber Lake Member seriously limit the degree to which the sandstone body can be reconstructed. Only its western edge is well defined, recent erosion has truncated its outcrop to the south and east, regional dip carries it underground to the north and northwest, and surficial deposits mask much of whatever remains east of the Missouri River. Within the type area of the Fox Hills Formation the few complete sections of the Timber Lake are insufficient for the construction of a very reliable isopach map. The thickness data available are generalized with other features of the Timber Lake sandstone on the map, Fig. 20. The maximum thickness of the Timber Lake Member measured in the type area is 150 feet; this section is incomplete at the top. The thickest complete section measured is 110 feet. Except along its western edge the sandstone appears to be consistently more than 50 feet thick. Further generalizations have little significance because of marked local variations in thickness. Although the bounding contacts diverge stratigraphically north and northeast beyond the type area the sandstone body itself is not a simple southwest-tapering wedge, instead the little that is known of the overall geographic distribution of its thickness indicates that the greater thicknesses do not coincide with the area of maximum stratigraphic range.

LITHOLOGIC CHARACTER

The Timber Lake sandstone varies locally in grain size, clay content, induration, type of bedding—or lack of it, kinds of concretionary masses, and in both the number and kind of its contained fossils. Considerable uniformity is found in the more obvious features of the unit, namely the weathered color, which is a dusky yellow to dark yellowish orange, and the characteristic concretionary masses that weather to dark values of red and yellow. The relatively yellowish hues of the Timber Lake sandstone contrast markedly with the grays of the underlying clays and clayey silts of the basal Fox Hills and Pierre Shale and with the paler brownish gray and light gray of the overlying Upper Fox Hills. Unweathered Timber Lake sandstone is dark gray to dark greenish gray.

The composition of the sand is relatively uniform provided that one accepts rather sharp local differences in the relative abundance of various accessory mineral grains as expectable in a sand unit deposited in a variety of environments and microenvironments that undoubtedly had differing energy patterns. In general the Timber Lake sand is a very fine-grained to medium-grained, dirty sand consisting of nearly equal parts of quartz and feldspar grains, a highly variable amount of clayey matrix, and an accessory mineral suite in which biotite, muscovite, epidote, and chloritic pseudomorphs of biotite are the more common constituents. Exclusive of the calcite and ferruginous cements of the concretionary masses, pyrite is the more ubiquitous secondary mineral; glauconite is generally present as scattered grains but tends to be a common constituent only in localized beds or intervals. The prevailing fashion is to call the Timber Lake sandstone a subgraywacke, which is acceptable if the term is used in its broadest compositional sense.

The grain size of the sand in the Timber Lake Member, illustrated in Fig. 8, falls predominantly in the very fine and fine grades of the Wentworth scale. Locally medium-grained sandstone dominates in some beds; much more rarely, small percentages of coarser grades up to very fine pebbles are found. A slight coarsening in grain size northward within the type area appears to take place in the Timber Lake sand body; as more fully explained below, this change is probably related to environment of deposition.

In addition to its sand-size constituents the Timber Lake sandstone contains highly variable amounts of interstitial silt and clay. Along its lateral facies boundaries, and at most places in its lower, clayey transitional phase, the relative percentage of the sand to the clay-silt fraction changes very gradually. At some places these different fractions are largely separated and thinly interbedded but generally they are mixed. Because the mixing is spotty and irregular, rather than homogenized, obvious edges to the sand body are rare and even the small-scale intertonguing that one might expect either is not present or not noticeable.

As a rule the fringe of clayey sandstone at its base and sides contains the highest percentage of the silt-clay fraction in the Timber Lake sand body. Within the upper part of the sand body there are local beds of clean, current-washed sandstone with little or no silt-clay fraction. The apparent limits of the sand body in the field are seldom the actual limits, for on the more deeply weathered exposures of gentle slopes the yellowish-orange coloration typical of the sand may extend well onto adjacent clayey silts, whereas in relatively fresh, steep exposures even sand with as little as 30 percent interstitial silt-clay weathers to much the same value of bluish gray as the underlying clayey silts. Between the extremes of clean sand and the clayey sand of the fringe areas a complete range of variation in the ratio of the silt-clay fraction to the sand is expectable and with careful selection of spot samples such a range in variation could no doubt be demonstrated. However the significant fact is that the high silt-clay content is usually found in the lower part and along the western edge of the Timber

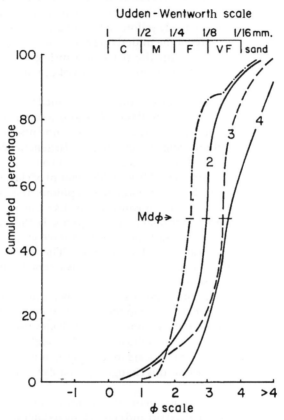

FIG. 8. Size composition of sands from the Timber Lake Member represented by cumulative curves.
1. Current-bedded sand of *Tancredia-Ophiomorpha* biofacies, Loc. 48, section 11, unit 7.
2. Sand with near-shore *Tancredia-Dosiniopsis-Ostrea-Pteria* association, Loc. 83.
3. Sand with abundant off-shore fauna including *Cucullaea*, Loc. 35, section 9, unit 5.
4. Clayey sand transitional to Irish Creek lithofacies, Loc. 35, section 9, unit 1.

Lake sandstone where it is associated with particular bedding structures and biofacies that are different from those in the remainder of the sandstone body, although to subdivide the Timber Lake sandstone body for discussion on the basis of its more clayey and less clayey parts would be an over-generalization obscuring its internal complexity.

The few size analyses of the Timber Lake sandstone indicate that the degree of sorting is locally variable, a fact that is quite evident from field inspection. The sand-size particles are generally well sorted as the range of dominant grades in the type area of the Fox Hills is small. Complications in sorting arise from two causes. Ferruginous aggregates are common and difficult to detect and eliminate from the sand; they tend to bias analyses toward the coarser grades. More significant changes in the degree of sorting result from the previously noted variation in the amount of mixing of the sand with the silt-clay fraction from place to place in the clayey peripheral areas of the sand body. Here the scarcity of fossils in the thinly interbedded parts is in marked contrast with the coincidence of the most varied and prolific invertebrate fossil assemblages with the massive or "mixed" parts of the transition. The abundant swirl structures and remnant fillings of borings in the massive beds, identical to those of the finer-grained "mixed" sediments of the underlying Trail City Member, indicate that the mixture of the sand fraction with the silt-clay fraction resulted largely from post-depositional churning by organisms rather than from conditions conducive to poor sorting during deposition of the sediment.

Composition of the Timber Lake sandstone is best studied in its calcareous concretions, which are ideal subjects for thin-section study of the constituents of the sand fraction as these can be seen preserved in the same condition as they were deposited. A pilot study by I. G. Speden of 15 thin sections of calcareous concretions from horizons throughout the Lower Fox Hills shows very little change in the clastic fraction throughout the concretions of both the Timber Lake and Trail City Members, indicating that the source remained the same during deposition of the Lower Fox Hills. Moreover, a single concretion taken from the Mobridge Member of the Pierre Shale about 220 feet below the base of the Fox Hills Formation shows the same relative proportions of major clastic constituents and the same suite of accessories as the Lower Fox Hills.

The thin sections show some kaolinization of feldspars, an abundance of chloritic pseudomorphs—many after biotite—as well as extensive chloritization of biotite, and, contrasting with these indications of chemical weathering, numerous angular grains of fresh feldspar. Although the few slides examined are insufficient to provide more than hints of possible directions for study, the contrast in fresh and weathered materials suggests a duality of source that the rising Laramide elements to the northwest and the then active Livingston volcanic field could readily have provided. The Timber Lake sand undoubtedly contains pyroclastic materials, glass shards are present in some of the thin sections and bentonite beds are locally preserved in the type area. Just northwest of the type area in the Linton region of Emmons County, North Dakota, the Timber Lake equivalent contains an unusual, thick, local deposit of little-altered volcanic ash (Stanton, 1917; Fisher, 1952; Manz, 1962; and others).

The sedimentary structures of the Timber Lake sandstone are as varied as its compositional features. Cross-stratification is more common than regular stratification, with tabular cross-laminated sets and lenticular cross-bedding and cross-lamination the dominant types. Regular stratification is limited chiefly to the thinly interbedded sand and silty clay found locally in the basal part of the Timber Lake, though, as previously noted, bedding in this part of the section has commonly been destroyed by the mixing action of burrowing organisms. Sedimentary structures other than bedding and traces of burrowing organisms are relatively rare. Some ripple marks are preserved in local ferruginous concretionary layers in the upper part of the Timber Lake sand and their apparent scarcity may result from the limitations of the poorly consolidated sand as a preserving medium. No desiccation structures of any kind have been observed.

Concretions are as conspicuous and characteristic of the Timber Lake Member as they are of the underlying Trail City Member, but they are more varied in kind and generally more abundant. Calcareous (calcitic) concretions, except for the obvious difference of having coarser detrital grains, are similar to those of the Trail City in most respects, ranging from punky to hard, jacketed to unjacketed, compressed to spheroidal, unfossiliferous to fossiliferous. Large concretionary masses, many as much as 5 or 6 feet in diameter, are more common in the Timber Lake Member. The calcareous, ovoid concretions are more common in the lower half of the Timber Lake, where they weather out in colors ranging from yellowish orange to dusty red and dusty brown.

Rusty-brown layers and irregular masses with both calcareous and ferruginous cement are abundant in the upper part of the Timber Lake Member and commonly form shelving ledges separated by poorly consolidated sand. Bright yellowish-orange, ferruginous impregnation of thin clay partings is also common and outcrops are locally sprinkled with the bright orange chips. The pustulate walls of the irregularly branching tubes called *Ophiomorpha* are everywhere preserved by ferruginous cementation.

STRATIGRAPHIC CONTROL

The columnar section (Fig. 7) illustrating key bed distribution in the Lower Fox Hills shows that a number of assemblage zones and bentonite beds occur in the section above the Trail City Member and its lateral equivalents. As previously noted, the bentonites are restricted chiefly to the Irish Creek lithofacies and occur only locally in the transitional clayey areas of the Timber Lake, very rarely extending any distance into the sand body itself. A second problem with key beds in the Timber Lake is that distinctive faunal associations of the assemblage zones are restricted to the southern part of the type area—the Moreau Valley and Cheyenne-Moreau divide —while northward a change of biofacies takes place, making it difficult to identify the less fossiliferous, faunally restricted extensions of the assemblage zones in the Grand River valley.

Although most of the marked changes in the aspect of the Timber Lake faunas takes place within the divide between the Moreau and Grand River valleys, the major zones of concretions in which the faunas occur are continuous from one valley to the other and can be matched with the aid of secondary control from the D bentonite, the widespread upper jarosite bed near the Trail City contact, and various key beds in the Trail City and equivalent Irish Creek lithofacies. In this way the stratigraphic details in at least the lower part of the sand body have been reliably linked between the Moreau and Grand Valleys. General lack of key beds in the upper part of the sand body does not permit as close control as in the lower part.

Key beds, assemblage zones and distinctive fossil associations of the Timber Lake lithofacies that are either helpful or confusing in the detailed correlation of its parts are described in ascending order in the paragraphs that follow.

C Bentonite: The C bentonite occurs peripherally to the Timber Lake sand body in the Irish Creek lithofacies west of the edge of the sand body in the Moreau Valley (Fig. 25) and also intermittently on the east end of the Cheyenne-Moreau divide east of the longitude of Goose Creek (Fig. 24). Stratigraphically it lies between the upper jarosite bed and the *Sphenodiscus* concretion layer. It has not been found outside of these two areas. In the area east of Goose Creek, where the beds above the upper jarosite thin appreciably, the bentonite is sandy, glauconitic and locally jarositic and may be merged with the upper jarosite for there is no evidence of the latter below the bentonite.

The *abyssinus* Concretions: The faunally distinctive concretions with *Discoscaphites abyssinus* which misleadingly occur at more than one horizon in the lower part of the Timber Lake and the upper part of the Trail City Member have already been described (p. 71).

Stratigraphic distribution of the concretions differs in the Moreau and Grand Valleys. In the latter they occur chiefly between the uppermost (*Nucula*) concretion layer in the Trail City Member and the upper jarosite. In the Timber Lake Member they occur only in and south of the Moreau Valley area where they are generally associated with either or both the *Sphenodiscus* concretion layer and the concretions of the *Cucullaea* Assemblage Zone. Where associated with the *Cucullaea* Assemblage Zone, the *abyssinus* concretions commonly are small and occur within an interval of a few feet just below the first abundantly fossiliferous concretions carrying the *Cucullaea* assemblage.

Sphenodiscus Concretion Layer(s): One, or two closely associated, persistent concretion layers are present between the upper jarosite and the *Cucullaea* Assemblage Zone. These are sparingly fossiliferous but in the Moreau Valley a considerable number of the concretions contain a single large specimen of *Sphenodiscus lenticularis*. The large nodose scaphitid *Discoscaphites nebrascensis* is also common, as are a variety of other invertebrates. *Sphenodiscus* concretions are generally large, ranging from a foot to as much as 5 feet in long diameter; they are hard, brittle, calcareous concretions that commonly weather deep reddish brown. Where *abyssinus* concretions are associated with the layer they occur most commonly just below its base. Where two concretion layers are present, which is true for much of the eastern and southern part of the type Fox Hills, it is the upper layer that generally is fossiliferous along the Moreau Valley.

Cucullaea Assemblage Zone: The most widespread and fossiliferous interval of concretions in the Timber Lake Member is the *Cucullaea* Assemblage Zone. It consists of an interval ranging from 5 to at least 20 feet thick which contains numerous red-brown-weathering concretions, scattered and in layers. A layer of large, closely spaced concretions near the base of the zone is a fairly persistent feature; these generally lie between 20 and 30 feet above the *Sphenodiscus* concretion layer. Many of the concretions contain a rich and abundant marine fauna. The zone is named for a large, thick-shelled bivalve *Cucullaea shumardi* Meek and Hayden, which is its most conspicuous species and occurs in both the concretions and loose in the surrounding sand.

Assemblages with *Cucullaea* are found throughout the Cheyenne-Moreau divide, along the Moreau Valley and northeastward to the easternmost point on the Moreau-Grand divide, but the diagnostic *Cucullaea* does not cross the midline of the Moreau-Grand divide and in the Grand Valley area the concretions in the interval contain a more sparse fauna in which *Pteria linguaeformis* and *Cymbophora warrenana* are the more conspicuous species. West of the Timber Lake lithofacies the concretions of the *Cucullaea* Assemblage Zone extend many miles into the Irish Creek lithofacies.

D Bentonite: The D bentonite, whose distribution in the Irish Creek lithofacies has been described (p. 78), can be traced eastward in the Grand Valley a short distance into the Timber Lake sand body. It is significant as one of a group of key beds that indicate the correlation of the *Cucullaea* Assemblage Zone, which it overlies in the Moreau Valley, with concretions bearing a sparser, *Cymbophora* association, which it overlies in the Grand Valley.

Cymbophora-Tellina Assemblage Zone: On the north side of the Moreau Valley between Irish Creek and Little Moreau River a fossiliferous zone of concretions occurs between 25 and 35 feet above the top of the *Cucullaea* Assemblage Zone. The ubiquitous *Pteria linguaeformis, Protocardia subquadrata* and *Ostrea pellucida* are the commoner elements but abundant *Cymbophora warrenana* (Meek and Hayden) and *Tellina scitula* Meek and Hayden give the fauna its distinctive appearance. Although more difficult to trace than other assemblage zones the *Cymbophora-Tellina* zone is significant for the control it furnishes along the west edge of the sand body as well as for the fact that it is the highest key bed in the Timber Lake sand that can be recognized with certainty in both the Moreau and Grand River valleys.

Tancredia-Ophiomorpha Sands: A characteristic feature of the Timber Lake lithofacies north of the midline of the Moreau-Grand divide are the beds of friable, light-rusty-brown sand and brown ferruginous concretionary layers, both of which commonly contain the burrow *Ophiomorpha* and locally contain beds or scattered specimens of the thick-shelled bivalve *Tancredia americana* Meek and Hayden. The sand is generally free of clay and silt and occurs in tabular, cross-laminated beds generally from 4 to 10 inches thick; the cross laminae commonly dip in opposite directions in immediately adjacent beds (see Pl. 5). This characteristic facies may locally be unfossiliferous, or contain only *Ophiomorpha*.

The faunally distinctive, current-bedded *Tancredia-Ophiomorpha* sands reach their greatest stratigraphic extent within the type area along the Grand River valley between Little Eagle and Bullhead where they occupy the entire Timber Lake sand body above the beds equivalent to the *Cucullaea* Assemblage Zone. Throughout the Grand River area it is the common biofacies of the upper half to two-thirds of the sand body. Both *Tancredia* and *Ophiomorpha* are found in the upper part of the Timber Lake lithofacies on the south side of the Moreau-Grand divide but here they are associated with a more prolific marine fauna above the level of the *Cymbophora-Tellina* assemblage. What might be called the "pure" *Tancredia-Ophiomorpha* biofacies does not seem to occur south of the midline of the divide.

INTERNAL FACIES AND THEIR DISTRIBUTION

The generalized isopachous map (Fig. 20) of the Timber Lake lithofacies reveals its overall geometry in the type area. The large sand body trends approximately north-northeast, terminates abruptly on its west side and thins gradually to the south and southeast. The original extent of the sand southward and southeastward is unknown; no deposits of it exist south of the Cheyenne-Moreau divide and none have been reported east of the Missouri south of the latitude of the mouth of the Grand River. The lower third of the Timber Lake can be demonstrated to thin appreciably southward and southeastward across the Cheyenne-Moreau divide as it passes into the Irish Creek lithofacies, and it is inferred that the entire sand body did not extend very far beyond its present outcrop in these directions. Strong supporting evidence for this is found in the distribution of Timber Lake fossils; the richest and most varied marine faunas are restricted to the outcrops along the Moreau Valley and on the Cheyenne-Moreau divide and it is only in these southern outcrops of the Timber Lake that off-shore marine faunas occur *throughout the entire thickness of the sand body.*

Lateral passage of much of the Trail City Member into the Timber Lake lithofacies to the north-northeast outside of the type area corroborates a southward direction of growth for the sand body. In Corson County north of the Grand River the outcrop area of the sand body extends eastward, appearing on both sides of the Missouri just south of the state line and in adjacent North Dakota. The impression from reconnaissance study is that the trend of the body swings more to the northeast in these areas but this may be more apparent than real, for younger beds largely obscure what happens to the Timber Lake lithofacies west of the Missouri River in North Dakota. It is important to bear in mind that in the type area of the Fox Hills Formation we are dealing with only part of the southern end—apparently the down-current, seaward end of a sand body whose size, extent and direction of trend to the north is essentially unknown.

The anatomy of the Timber Lake sand body is complex and its local lithologic and faunal variations appear disordered on first acquaintance, but when their geographic distribution is plotted at successive levels, intelligible patterns emerge. The following account is organized around the series of distribution maps (Figs. 21 to 23) constructed at the primary control levels in the Timber Lake sand, the *Sphenodiscus* concretion layer, the concretions of the *Cucullaea* Assemblage Zone, and those of the *Cymbophora-Tellina* Assemblage Zone. As was true of concretion zones in the underlying Trail City Member, the three Timber Lake concretion zones persist through changes in their fossil assemblages and in the lithology of surrounding beds, but where they pass laterally into the *Tancredia-Ophiomorpha* sands they commonly disappear or cannot be identified with confidence. As all three horizons can be identified with reasonable certainty on both the east and west sides of the sand body such local disappearances into the center have little effect on the construction of the distribution maps. The greatest possibility for error is in north-south reconstructions across the Moreau-Grand divide where the major faunal changes take place and where neither continuous tracing nor close spacing of adequate sections is possible. Reliance here is mostly on matching sections across the divide. The results are considered sound because the number and spacing of concretion zones in corresponding parts of the sand body is approximately the same in both the Grand and Moreau Valleys and the zones occur at about the same distance above the persistent upper jarosite bed. On the west side of the sand body where the greatest gap between measured sections occurs, the D bentonite holds the same position relative to the concretion zones in both valleys, facilitating correlation across the divide.

LEVEL OF SPHENODISCUS CONCRETIONS

Disregarding the sandy zones of the Trail City Member that are apparently early tongues of the Timber Lake sand body and contiguous with it to the north outside the type area, the first appearance of an appreciable amount of sand is in beds adjacent to the upper jarosite layer. At this horizon the distribution of facies is obscure as the sand is in an irregularly clayey transition phase—here more sandy, there more clayey— over most of the area. Lateral change into the Irish Creek lithofacies is very gradual and broad patterns are further obscured by local variations in the thickness and character of the key jarosite, locally glauconite, itself. The few well-defined distributional features at this horizon are not sufficiently different from the facies distribution at the level of the *Sphenodiscus* concretion layer, 15 to 30 feet higher in the section, to justify a separate map.

Facies distribution at the level of the *Sphenodiscus* concretions, shown in Fig. 21, is representative of the early, transitional phase of the Timber Lake sand body in the type area. Typical Timber Lake sand with only minor clay content is limited to a narrow, elongate lens extending into the area from the northeast, crossing the Grand River around Little Eagle and terminating an unknown distance to the southwest under the Moreau-Grand divide. Peripheral to this sand the Timber Lake consists of appreciably more clayey sand divisible locally into separate facies on the basis of distinctive lithologic or faunal characteristics. Along the west side of the type area this clayey sand grades into the Irish Creek lithofacies.

Sandstone Facies: The dusky yellow sandstone containing the concretions of the *Sphenodiscus* level is best exposed in the bluffs north of the Grand River around Little Eagle; here it is chiefly massive but locally contains scattered, thin, rusty-weathering clay partings. The concretions are brown- to red-brown-weathering sandy limestone of various shapes, a very few of which locally contain a restricted marine fauna consisting chiefly of a *Pteria linguaeformis-Ostrea pellucida* association, less commonly with *Dosiniopsis* and several genera of gastropods.

In these bluffs north of Little Eagle the sand is continuous downward below the horizon of the key jarosite, extending to about 10 feet above the *Nucula* concretion layer in the Trail City Member—the lowest level at which the Timber Lake litho-facies is found in or south of the Grand River valley. Westward, southward and east-ward it rises markedly in the section but northward and northeastward toward the Missouri it drops progressively lower in the section. The limits of the sandstone shown in Fig. 21 were drawn where the base of the sand is estimated to lie approximately at the level of the lowest concretions in the *Sphenodiscus* layers.

Northeast of Oak Creek neither exposures nor stratigraphic control are adequate but it appears that the sandstone, at least locally, has the *Tancredia-Ophiomorpha* biofacies; the associated sedimentary features include tabular, cross-laminated bed-ding, more orange hues of weathered sand and shaley partings, and irregular, ledgy ferruginous concretion layers.

Rock Creek Lithofacies: West along the Grand River from the Little Eagle area the sandstone facies grades to a distinctive thin-bedded sand and shale, here informally named the Rock Creek lithofacies for plentiful exposures in bluffs along lower Rock

Creek and around its confluence with the Grand River at Bullhead. Local thin-bedded intervals are not uncommon anywhere in the transitional facies of the Timber Lake, but only in the Rock Creek lithofacies is the bedding sharply defined and vertically persistent. Outcrops of the Rock Creek lithofacies are noticeably banded (Pl. 4, fig. C) and, because of the relatively high sand content, weather yellowish gray to tan. Details of its lithology are given in the reference section on p. 113.

Concretions of the *Sphenodiscus* level are not traceable into the Rock Creek lithofacies, which contains chiefly small, flattened, dusky-red-brown concretions that appear scattered rather than concentrated in definite layers. The approximate position of the level can be traced using its distance above the upper jarosite, which is fairly persistent along the Grand River west of the Little Eagle area.

At its greatest stratigraphic interval, the Rock Creek lithofacies extends from just above the *Nucula* concretion layer of the Trail City to just below the level of the *Cucullaea* Assemblage Zone. In the Rock Creek area thin-bedding dies out down-ward, disappearing approximately at or as much as 10 feet above the upper jarosite and giving way to mixed, glauconitic, sandy clay. The latter locally contains loose shells of *Ostrea pellucida*, *Dosiniopsis*, small *Belemnitella* and a variety of gastropods above the jarosite, and fossiliferous, punky concretionary masses that include *Pteria linguaeformis*, *O. pellucida*, *Protocardia* and *Discoscaphites abyssinus* below it. Fossils are rare in the thin-bedded layers and occur chiefly as isolated shells near the base and just below a more massive sand, at the level of the *Cucullaea* zone, into which the Rock Creek lithofacies grades at the top. A small fauna characterized by *Dosiniopsis* and *Ostrea pellucida* is the principal association found. Rarely this same association also occurs within the Rock Creek beds.

The westward change from the Rock Creek thin-bedded lithofacies to the Irish Creek lithofacies is gradational but fairly abrupt. The decrease in sand and increase in clay results in a change of color on the outcrop to gray and dark gray. The thin-bedding becomes more localized and less obvious; much of it is replaced by massive-appearing mixed sand and clay and, in what remains, the sand layers are thinner and more irregular. Along the Grand River this change to the Irish Creek lithofacies takes place within 2 or 3 miles east of the mouth of Hump Creek, and with the change a persistent concretion layer reappears at the *Sphenodiscus* level. Change of the Rock Creek thin-bedded lithofacies to predominately massive sandy clay above the con-cretions of the *Sphenodiscus* level and to silty gray clay with irregular thin beds of sand and silt below the concretion layer is a pattern that persists from the area of gradation into the Irish Creek lithofacies.

Outside of the Grand River area the Rock Creek lithofacies is rarely exposed. Outcrops of the interval along Firesteel Creek are either in the Irish Creek lithofacies or well along in the gradation to it. Outcrops along Oak Creek between 4 and 5 miles southwest of McLaughlin show that at least the beds beneath the level of *Sphenodis-cus* concretions are in the Rock Creek lithofacies, but beds at and above this level appear to be in sand with a *Tancredia-Ophiomorpha* biofacies.

Clayey Sand Facies: South of the Grand River valley the beds at and adjacent to the *Sphenodiscus* concretion layer are best treated as a single somewhat variable, facies. This consists chiefly of mixed sand and clay in varying amounts but with sand generally predominant. Obscure thin-bedding is locally present but more commonly the structure is massive and shows blebs and swirls of clay in the sandier parts and of sand in the clayey parts. A persistent layer of red-brown-weathering concretions which vary in size from a foot to 4 feet in long diameter forms the *Sphenodiscus* concretion layer. Locally a second layer and/or scattered smaller concretions occur within 10 feet below it.

Along the Moreau Valley the clayey sand facies grades westward into the Irish Creek lithofacies in the vicinity of the Dewey-Ziebach County line. As there is no marked change in internal structure the gradation is evident chiefly by a change from clayey sand to silty or sandy clay and the corresponding change in weathered color from light yellowish or brownish gray to gray. The corresponding change along the western part of the south edge of the Cheyenne-Moreau divide is obscured by poor exposures. A more clayey phase is also apparent in the few exposures on the east ends of both the Moreau-Grand and the Cheyenne-Moreau divides. Here, intervals of gray silty or sandy clay are interbedded with the clayey sand facies and appear to thicken south-eastward at the expense of the latter. Complete change to the Irish Creek lithofacies does not occur anywhere on the east ends of the divides in the beds above the upper jarosite.

The clayey sand facies not only becomes more clayey around its periphery, it also is visibly more sandy in the bluffs of the Moreau and its tributaries in the area between the mouths of Red Earth and Redwater Creeks. In parts of this general area the interval including the *Sphenodiscus* concretion layer is in a sand with only scattered clay blebs and little interstitial clay. No clear-cut sand body can be distinguished, however, for the sand is scattered in lenses within clayey sand. The apparent concentration of sandier lenses in this limited area and the gradual increase in clay laterally from it, both east and west along the Moreau Valley, indicates the approximate position of the axis of the Timber Lake sand body. This suggests that beneath the younger beds on the Moreau-Grand divide the sandstone facies of the Grand River valley changes to a more southerly trend as it loses its identity southward in the clayey sand facies.

The *Sphenodiscus* concretions of the clayey sand facies are commonly fossiliferous in the Cheyenne-Moreau divide west of Goose Creek and in the area west of Little Moreau River north of the Moreau. Throughout these areas concretions containing single large specimens of *Sphenodiscus lenticularis*, some as much as 16 inches in diameter, are common enough to typify the layer (Pl. 7, fig. A). Immediately below this layer are the more abundantly fossiliferous *abyssinus* concretions characterized by scaphitids, including many juveniles, *Pteria linguaeformis*, *Oxytoma nebrascana*, *Ostrea pellucida*, and the protobranchs *Yoldia evansi* Meek and Hayden and *Y. scitula* Meek and Hayden. *Discoscaphites nebrascensis* commonly occurs with *Sphenodiscus*, and the bivalve *Cucullaea nebrascensis* Owen is locally present and becomes increasingly more abundant as this level is followed southward into the Cheyenne-Moreau divide.

Eastward and northeastward from their abundantly fossiliferous area the *Sphenodiscus* concretions contain a sparse and less varied fauna consisting chiefly of *P. linguaeformis*, *O. pellucida* and scaphitids. Westward the *Sphenodiscus* concretions extend well into the Irish Creek lithofacies but beyond a few miles west of the edge of the clayey sand facies they are generally unfossiliferous.

LEVEL OF THE CUCULLAEA ASSEMBLAGE ZONE

Geographic distribution of local biofacies and lithofacies within the Timber Lake sand body at the level of the *Cucullaea* Assemblage Zone is shown in Fig. 22. This is the more reliable of the Timber Lake distribution maps because of the ease with which the abundant fossiliferous concretions of the assemblage zone can be identified and traced. Concretions at this level are absent only in the sands with the *Tancredia-Ophiomorpha* fauna and in the adjacent area to the west where sands transitional to the latter are present. Fossiliferous concretions at the *Cucullaea* level extend farther into the Irish Creek lithofacies than do those at any other level in the Fox Hills Formation (p. 75).

The facies pattern within the sand body differs from that at the level of *Sphenodiscus* concretions chiefly in 1) the broader extent of the current-bedded sands with the *Tancredia-Ophiomorpha* association, which have spread southward and westward, 2) the absence of the Rock Creek lithofacies, and 3) the decrease in clay content of the massive sands as a whole.

The most abundant and diverse faunas in the Timber Lake sand body occur at the level of the *Cucullaea* Assemblage Zone. They are differentiated into several distinctive biofacies so distributed geographically that fossil assemblages south of the Moreau-Grand divide are superficially quite distinct from those north of the divide. A varied marine fauna with *Cucullaea* conspicuously present occupies the southern outcrops, whereas sparser, less diverse faunas lacking *Cucullaea* occupy the outcrops along the Grand River valley. The latter include the distinctive *Tancredia-Ophiomorpha* biofacies.

As in its lower part, the Timber Lake sand body terminates relatively sharply along its west side by abrupt gradation into the Irish Creek lithofacies. This is marked by a conspicuous change on bare, weathered outcrops from yellowish-orange, rounded bluffs to gray, steeply concave, clayey slopes. The western edge of the sand body south of the Moreau-Grand divide is approximately in the same position as at the level of the *Sphenodiscus* concretion layer, but in the Grand River valley it has shifted several miles westward. On the east side of the sand body, gradation to very clayey sand is evident in the few exposures on the ends of both divides. On the Cheyenne-Moreau divide this change begins at least 14 miles west of the easternmost outcrop; on the Moreau-Grand divide it probably begins about 8 or 10 miles from the east end of the divide. None of the eastern exposures show a complete change to the Irish Creek lithofacies, indicating a considerably more gradual lithofacies change than on the west side of the sand body.

Cucullaea Biofacies: The *Cucullaea* Assemblage Zone extends throughout the Timber Lake and Irish Creek lithofacies of the Fox Hills Formation over more than half of the type area (Fig. 22). As previously noted, this zone is an interval of varying thickness containing many concretions which are abundantly fossiliferous. Although the greatest number and variety of fossils in this rich molluscan assemblage are confined largely to this zone, *Cucullaea* itself locally ranges both above and below it in certain parts of the Timber Lake sand body; consequently it is expedient to recognize a *Cucullaea* biofacies as distinct from, but including, the *Cucullaea* Assemblage Zone. With rare exceptions the first appearance of *Cucullaea* is in the Timber Lake sand body. The exceptions consist of a few specimens all of which were found in local lenses of glauconitic, clayey sand in or near the upper jarosite bed at four scattered localities (121, 196, 210, 267) in the Moreau and Grand Valleys. Within the Timber Lake sand body, *Cucullaea* and a few other species commonly occur, free or in concretions, in the massive sand above the concretions of the assemblage zone. At a few localities in the Cheyenne-Moreau divide, *Cucullaea* has also been found in sand and slightly clayey sand as much as 15 feet beneath the basal concretions of the assemblage zone.

The maximum stratigraphic extent of *Cucullaea* in the Timber Lake sand occurs in the Moreau Valley outcrops and on the Cheyenne-Moreau divide, west of the mouth of Redwater Creek. North of the Moreau in this area, 20 to 30 feet of the sequence contains *Cucullaea* and this range increases south of the river into the divide where as much as 55 feet of beds containing *Cucullaea* are known (Loc. 35). This latter area of abundant *Cucullaea* on the Cheyenne-Moreau divide is the most fossiliferous area of the Timber Lake sand in the type Fox Hills. Although the rich marine molluscan fauna is concentrated in the assemblage zones, fossils occur almost continuously throughout the Timber Lake sand and *Cucullaea* distribution appears to extend from below the *Cucullaea* concretions to the level of the *Cymbophora-Tellina* Assemblage Zone. This abundance of fossils along the southwest side of the sand body makes the abrupt change of the Irish Creek lithofacies all the more striking. Only the fossiliferous concretions of the *Cucullaea* Assemblage Zone itself extend into the Irish Creek, where both the concretions and the specimens of *Cucullaea* become smaller and fossils are largely restricted to the concretions (Pl. 6, fig. B).

Exposures of the Timber Lake east of S. Dak. Highway 63 in the Moreau Valley and on the Cheyenne-Moreau divide are generally poor but indications are that *Cucullaea* is restricted chiefly to its assemblage zone and not more than about 10 or 15 feet of overlying beds.

Reduced Biofacies of the Grand River Area: Concretions at the level of the *Cucullaea* Assemblage Zone in the eastern and western outcrops of the Timber Lake sand body along the Grand River valley contain much less diverse and abundant faunas in which *Cucullaea* is not present. These two areas are separated by Timber Lake sand carrying the *Tancredia-Ophiomorpha* biofacies. In both the eastern and western faunas *Pteria linguaeformis, Cymbophora warrenana, Tellina scitula, Protocardia subquadrata* and *Ostrea pellucida* are common, the first three being the more conspicuous elements. Of these only *Cymbophora* is quite rare in the fauna containing *Cucullaea* to the south, and so its presence, along with the absence of *Cucullaea,* is the most obvious difference between the northern and southern faunas of the *Cucullaea* Assemblage Zone. The fauna is referred to hereafter as the *Cymbophora-Tellina* association, or fauna.

Although the eastern and western *Cymbophora-Tellina* faunas in the Grand River area are similar the western fauna commonly also contains *Tancredia* and, more rarely, *Panope.* Neither of these genera has yet been found at this level in the eastern part of the Grand Valley. The western fauna extends into the Irish Creek lithofacies with little apparent change except *Tancredia* becoming as common as *Cymbophora.*

Gastropods and cephalopods are common in both the eastern and western faunas but neither group is as abundant or as diversified as in the southern assemblages with

Cucullaea. Perhaps the most striking difference is the rarity of *Sphenodiscus* in the western *Cymbophora-Tellina* faunas of both the Timber Lake and Irish Creek lithofacies.

Tancredia-Ophiomorpha Biofacies: Current-bedded sands with *Tancredia* and *Ophiomorpha* extend southwestward across the Grand River, between Little Eagle and Bullhead, at the horizon of the *Cucullaea* Assemblage Zone. These beds disappear under younger Timber Lake beds south of the Grand and terminate under the higher part of the Moreau-Grand divide, for where the level reappears again south of the divide the rocks are in the *Cucullaea* biofacies. In the Grand River valley two distinct but gradational phases divide the *Tancredia*-bearing sands into eastern and western parts (Fig. 22). The somewhat larger eastern part contains a typical *Tancredia-Ophiomorpha* assemblage in rusty-weathering, relatively clean, fine-grained sand with tabular, cross-laminated bedding. *Ophiomorpha* is relatively common throughout, while *Tancredia* is restricted largely to localized accumulations with very few other shells except the gastropod *Lunatia*.

Locally in the area just east of the longitude of Bullhead this biofacies grades west into one in which *Ophiomorpha* and *Tancredia* occur together with a more varied fauna including *Dosiniopsis* and *Ostrea pellucida*. Unlike the typical *Tancredia-Ophiomorpha* sands, the beds with the *Dosiniopsis* association commonly appear massive or show some lamination and concretion layers reappear. Just west of Bullhead this transitional fauna grades rather abruptly into the western *Cymbophora-Tellina* association. *Dosiniopsis* associations in the vicinity of Bullhead were mentioned in the discussion of the Rock Creek lithofacies (p. 101); they are more common in this area than elsewhere in the type area of the Fox Hills. Their occurrence in and adjacent to the margins of the Rock Creek lithofacies just west of sands with the restricted *Tancredia-Ophiomorpha* biofacies indicates that they are a persistent, if not conspicuous, transitional phase of the latter.

LEVEL OF THE CYMBOPHORA-TELLINA ASSEMBLAGE ZONE

The beds containing the fauna for which this thin interval of fossiliferous concretions as named crop out chiefly on the north side of the Moreau River between Irish Creek and Redwater Creek. Beds at this level are largely limited to the higher, less well-exposed, parts of the divides, and good exposures are found mostly in the limited area of river bluffs where the outcrop crosses the valleys. As a result the patterns shown on the map (Fig. 23) are necessarily sketchy.

The principal features of the sand body at this horizon are 1) the domination of the northern part of the type area by sands containing the *Tancredia-Ophiomorpha* biofacies, 2) the spread southward of a *Cymbophora-Tellina* biofacies in the area previously occupied by the *Cucullaea* biofacies, 3) the appearance of the Bullhead lithofacies of the Iron Lightning Member in place of the Irish Creek lithofacies west of the sand body. Note also that at this horizon in the north the western margin of the Timber Lake sand body has shifted farther westward so that it is trending nearly north-south.

Moreau Valley and Cheyenne-Moreau divide: Concretions of the *Cymbophora-Tellina* Assemblage Zone form a continuous layer high in the Moreau Valley bluffs on either side of S. Dak. Highway 65, and can be followed on the tributary divides for several miles east of the highway on the north side of the river. West of the highway, abrupt facies changes at this level, first to Irish Creek lithofacies then to the Bullhead lithofacies, eliminate the assemblage zone within three miles—its last good exposures being along the heads of draws at the foot of St. Patrick's (Ragged) Butte (Loc. 56). Tracing the *Cymbophora-Tellina* Assemblage Zone along the north side of the Moreau is very difficult because the level passes into grass-covered divides and exposures of it are rare. Locally, fossiliferous concretions occur in a highly glauconitic zone at this horizon in the vicinity of Meadow Creek (Loc. 34 and 38), where *Pteria* and *Protocardia* were the only bivalves found. Similar glauconitic sands exposed along S. Dak. Highway 63 (Loc. 42) contain scattered *Pteria* and ammonoid fragments, and associated ferruginous concretion layers carry a few *Ophiomorpha*. In contrast, richly fossiliferous concretions with the *Cymbophora-Tellina* association were found by Speden 5 miles east of Highway 63 (Loc. 122) in the road bed of a prairie trail at approximately the altitude that the assemblage zone should occur in this area. East of this a single occurrence of the zone was found on the easternmost tip of the Moreau-Grand divide, 5.5 miles due east of Trail City (Loc. 111). Here punky, semi-indurated sand masses with fossil clusters contain the key fauna and a variety of other species. Between these two last, widely spaced localities the scarce outcrops of timber Lake sand are chiefly current-bedded and have scattered *Ophiomorpha*.

South of the Moreau River the *Cymbophora-Tellina* Assemblage Zone has been identified with certainty only in the vicinity of S. Dak. Highway 65 at the top of the bluffs near the river. Elsewhere on the Cheyenne-Moreau divide, areas east of the change to the Bullhead lithofacies and high enough to include the *Cymbophora-Tellina* zone generally underlie grassy prairie. Here scattered small exposures of the Timber Lake sand commonly contain *Cucullaea,* but because of the stratigraphic range of the latter in this area the level of most of these cannot be ascertained. In the thickest single exposure of the Timber Lake, measured on the divide near Lantry (Loc. 35), 55 feet of sand contains *Cucullaea,* loose and in concretions, throughout all but the uppermost 3 or 4 feet. While it is possible that concretions of the *Cymbophora-Tellina* Assemblage Zone have not been detected because of poor exposures in their limited area of potential outcrop, it is more likely that the fauna at this level has changed to a *Cucullaea* association south of the Moreau.

Several pieces of indirect evidence suggest that the latter possibility is the more probable. First, the distribution of associations throughout the type area at the underlying *Cucullaea* Assemblage Zone level reveals that the pattern of change from a biofacies with *Cucullaea* on the south to one with *Cymbophora-Tellina* on the north existed previously on the Timber Lake sand body. Southward shift of biofacies, however, could have brought about either of the two possible patterns, depending on the extent of the shift. Rare occurrence of individual concretions with *Cucullaea*—and without *Cymbophora*—in and just below the *Cymbophora-Tellina* Assemblage Zone on the north side of the Moreau at several places (Loc. 34 - see Section 8, unit 9, and Locs. 210, 212) strongly suggest the beginning of a biofacies change in the Moreau Valley area. Such a change would also help explain the 55 feet of beds with *Cucullaea* at Lantry, the thickest interval with *Cucullaea* known in the type area and thick enough to encompass both the *Cucullaea* and *Cymbophora-Tellina* Assemblage Zones, judging from their interval of separation in the Timber Lake sand north of the Moreau. On the strength of these indications and the lack of evidence to the contrary this interpretation is adopted for the map (Fig. 23); the inferred extent of the *Cucullaea* biofacies eastward is based largely on topography.

Grand River Area: The *Tancredia-Ophiomorpha* biofacies occupies most of the outcrop at the *Cymbophora-Tellina* level in the Grand River area, extending westward to the mouth of Hump Creek. To the west along the Grand and in the contiguous valley of Firesteel Creek, massive sands occur that contain a widespread fossiliferous interval, including persistent concretions, at a horizon comparable to the *Cymbophora-Tellina* Assemblage Zone in the Moreau Valley, lying between 25 and 35 feet above the fossil assemblages equivalent to the *Cucullaea* Assemblage Zone. The fauna consists chiefly of *Ostrea pellucida*, *Pteria linguaeformis* and *Tancredia americana* which occur loose in the sand and in punky concretions built around clusters of *O.pellucida*. Large, tabular to ovoid, red-brown-weathering limestone concretions mark the base of the fossiliferous interval. Plant fragments are abundant throughout and the large shells of *Panope occidentalis* Meek and Hayden are commonly found in life position in the sand. Farther west along the Grand River as the sand becomes clayey and grades into the Irish Creek lithofacies, scattered concretions at this horizon locally contain *Cymbophora* and *Protocardia* as common elements in addition to *Ostrea*, *Pteria* and *Protocardia*.

The concretions associated with the *Cymbophora-Tellina* level along Firesteel Creek and the area around its confluence with the Grand River can be traced eastward along the Grand into the *Tancredia-Ophiomorpha* biofacies. The layer becomes discontinuous but concretions occur at its level at scattered localities as far east as Bullhead. These concretions are commonly associated with massive to thinly-bedded sands that locally carry a fauna in which *Ostrea*, *Pteria*, *Dosiniopsis* and *Tancredia* are the commoner elements. Again a transition facies with *Dosiniopsis* flanks the more typical *Tancredia-Ophiomorpha* biofacies in the vicinity of Bullhead. East of Bullhead only the latter has been found at the horizon in question.

TIMBER LAKE SAND ABOVE THE CYMBOPHORA-TELLINA LEVEL

The upper part of the Timber Lake sand body in the type area may contain well-defined, widespread levels with distinctive fossil associations but if so they are obscured because of the general lack of exposures. The westward-rising land surface on the divides barely attains the critical level for preserving the upper sand before the western limit of the sand body is reached. Apparently none of it remains on the Cheyenne-Moreau divide. On the south part of the Moreau-Grand divide between Red Earth and Redwater Creeks there is good indication in scattered buttes that another abundantly fossiliferous zone is present 30 or 40 feet above the *Cymbophora-Tellina* level, but the horizon cannot be traced into the Grand River area. In the latter area the *Tancredia-Ophiomorpha* biofacies dominates the upper part of the sand body.

Along its western edge the upper part of the Timber Lake sand body changes abruptly to the Bullhead lithofacies of the Iron Lightning Member. At about 25 feet above the level of the *Cymbophora-Tellina* zone a shift eastward from 2 to 10 miles is evident in the position of the western edge of the sand body, the greater distance being in the Firesteel Creek—Grand River area where the edge of the body returns approximately to the position it had at the level of *Sphenodiscus* concretions.

The greatest thickness of sand above the *Cymbophora-Tellina* level—50 feet—was measured on the axis of the sand body just southwest of Bullhead (Loc. 253); the top of the section here was short of the contact with the overlying Bullhead lithofacies. Throughout all but the marginal area of the sand body the upper part was probably at least 40 feet thick.

Tancredia-Ophiomorpha Biofacies: Along the Grand River there is little significant change in the character of the sand body. Beds with the *Tancredia-Ophiomorpha* biofacies are essentially continuous except for zones of thin-bedded to laminated sand which tend to be more conspicuous west of Bullhead. One such zone, more shaly than the others, lies between 25 and 35 feet above the *Cymbophora-Tellina* horizon and appears to be fairly persistent. It roughly divides the upper part into two units of *Tancredia-Ophiomorpha* sand.

On the west side of the sand body along Firesteel Creek and the western Grand River area, where the *Cymbophora-Tellina* Assemblage Zone carries an *Ostrea-Pteria-Tancredia* association, the overlying sand contains chiefly *Tancredia* and *Ophiomorpha*. But at two localities (48, 272) the upper 4 to 10 feet of sand beneath the contact with the Bullhead lithofacies contains *Ostrea-Pteria* clusters; at locality 272 these are associated with *Cymbophora, Tancredia,* and *Panope*. The two localities may not be at the same horizon, for the position of the base of the Iron Lightning Member varies in this border area of the sand body, but the record of a return to conditions favoring the *Ostrea-Pteria* association just before the lithofacies change to the Iron Lightning is noteworthy. A similar fauna occurs locally at the top of the Timber Lake north of the town of Bullhead (Loc. 84), just beneath the Iron Lightning contact.

Tancredia-Ophiomorpha Biofacies Equivalents: Where best represented, along the west edge of the sand body in the Moreau Valley area, the *Cymbophora-Tellina* Assemblage Zone lies within a few feet of the top of the Timber Lake lithofacies, which is succeeded by the Bullhead lithofacies of the Iron Lightning Member. Seven miles to the east and throughout much of the southern part of the Moreau-Grand divide east of R. 22 E. the zone, or its projected interval, is overlain by at least 40 feet of sandstone. At most places this sandstone is friable and exposures of it are few, but locally it is secondarily indurated and forms the caps of numerous buttes along the north side of the Moreau. The butte caps have in the past been called Colgate but the sand which forms most of them is part of the Timber Lake sand body. The relationship of this with sand of similar lithology in the Upper Fox Hills is discussed in more detail in another place (p. 129); here we are concerned primarily with its fauna. Road cuts and buttes for 6 or 7 miles on either side of S. Dak. Highway 63 north of the Moreau River breaks show small exposures of the sandstone and many of these contain a fauna consisting predominantly of *Ostrea pellucida* and *Pteria linguaeformis*. *Tancredia americana* and *Cymbophora warrenana* are common and *Ophiomorpha* is generally present but not usually in abundance. *Panope* and *Tellina* and the ammonoids *Discoscaphites nebrascensis, D. cheyennensis* and *D. abyssinus* are found locally. The most characteristic feature of the fauna is the presence of large numbers of *O. pellucida* and *P. linguaeformis* in dense clusters 4 to 10 inches in diameter. These may include an occasional ammonite or other shell but *Tancredia* and *Cymbophora* occur chiefly as scattered shells or in patches on bedding planes.

Because of the discontinuous, small outcrops the thickness of the interval occupied by this fauna is not exactly known nor is the distance of its first appearance above the *Cymbophora-Tellina* Assemblage Zone below. The interval is obviously a lateral equivalent of both the *Tancredia-Ophiomorpha* biofacies of the central Grand River area and of the Iron Lightning Member to the west. Whether the fossils in the sand are restricted to definite horizons cannot be demonstrated; at one locality (281) two zones of scattered *Ostrea-Pteria* clusters, with associated *Ophiomor_pha* in the surrounding sand, are separated by 30 feet of sand with scattered *Ophiomorpha*. But in general the impression from the scattered outcrops is that the faunas occur intermittently in lenses throughout the sand from 20 feet above the *Cymbophora-Tellina* horizon to the top of the Timber Lake exposures.

East of Red Earth Creek along the north side of the Moreau, the nature of the upper contact of the Timber Lake sand is obscure. The sands above the *Cymbophora-Tellina* zone apparently are thickest between Red Earth Creek and S. Dak. Highway 63, across the thicker axial part of the sand body as a whole (Fig. 20). Locally in this area, lignitic shale, lignite and local patches of salmon-pink, baked shale, or "red dog", crop out within 10 feet of the *Ostrea-Pteria* sands and no Iron Lightning Member is in evidence. No exposure showing a clearcut contact was seen, but it is evident that the top of the sand in question is locally very close to, and probably in contact with, the basal lignite of the Hell Creek Formation. Apparently most of the Iron Lightning sequence grades into the Timber Lake sand on the axial region of the sand body in this area. The same stratigraphic relationship may obtain along the axis of the sand body on the north side of the Moreau-Grand divide around the narrow area of Hell Creek outcrop east of Firesteel Creek. Here too, outcrops of the Iron Lightning Member are exceedingly rare and many of the so-called "Colgate" sand exposures have *Ostrea-Pteria* clusters and are continuous downward into Timber Lake sand.

SUMMARY OF FACIES PATTERNS

Changes in fossil associations, chiefly among bivalve species, permit a more definitive subdivision of the Timber Lake sand body than do changes in lithology, although the latter are also useful. The lithofacies gradient is from fine-grained sand to finely sandy clay and most of the more abundant bivalves used to distinguish biofacies apparently lived in, or on, most parts of this range of bottom sediment, except where current-bedded sand indicates a turbulent environment. Factors other than type of bottom sediment were apparently responsible for most of the biofacies distribution.

The biofacies also form a gradient with the species-poor *Tancredia-Ophiomorpha* association at one end and the diversified association with *Cucullaea* at the other. Between these are three conspicuous gradational associations that recur at different levels and places in the Timber Lake sand body; their principal features are summarized below.

1) A *Tancredia-Dosiniopsis* association with abundant *Pteria* and *Ostrea* occurs locally, adjacent to the *Tancredia-Ophiomorpha* association on the west, in the Grand River area at and above the level of the *Cucullaea* Assemblage Zone. A closely related association in which *Tancredia* is rare or absent is found at a few localities in the underlying Rock Creek thin-bedded lithofacies in this area.

2) A *Tancredia-Cymbophora* association with abundant *Pteria* and *Ostrea* is widespread at several levels and is the principal association with *Tancredia* outside of

the faunally restricted *Tancredia-Ophiomorpha* biofacies. It occurs in the relatively clean, fine- to medium-grained sands in the top of the sand body north of the Moreau River and also in the clayey sands transitional to the Irish Creek lithofacies at the *Cucullaea* Assemblage Zone level along Firesteel Creek.

3) A *Cymbophora-Tellina* association with abundant *Protocardia*, *Pteria* and *Ostrea* is widespread and occurs, geographically, between the *Cucullaea* associations and the associations with *Tancredia*. The sediment in which it occurs ranges from slightly clayey, fine-grained sand to sandy clay.

As the maps at the different levels of the sand body indicate, the *Tancredia-Ophiomorpha* biofacies in current-bedded sand forms an axial element of the sand body that spreads southward and laterally with its growth. The total advance of the south edge of this facies in the type area of the Fox Hills is between 30 and 40 miles. The faunally more diverse *Tancredia*-bearing biofacies border this axial facies on the west and south. *Tancredia*-bearing biofacies do not extend into the Moreau River area until the final stage in the development of the sand body. The *Cymbophora-Tellina* association appears at the level of the *Cucullaea* Assemblage Zone and both here and above occurs geographically between the *Cucullaea* associations and those with *Tancredia*. Southward shift of these associations in concert with the *Tancredia* associations is apparent.

Both the lithofacies and biofacies patterns within the Timber Lake indicate a sand body growing southward under a current from the north or northeast. The body is asymmetrical, being steeper on its western, shoreward side. Rich shallow-water marine faunas occupied its advancing seaward end and may have extended northeastward along its seaward side. Less diverse faunas inhabited the shallow parts on either side of its axis. The shallowest, axial part of the body supported a fauna whose low diversity probably reflects the turbulent environment indicated by the current-bedded sands.

REFERENCE SECTIONS

The following sections illustrate most of the major aspects of the Timber Lake Member. The impossibility of selecting any one section of this variable sand body as typical is obvious. Section 8, below, is chosen as the principal reference section to compensate for the lack of an originally designated type section. It is more complete than any other exposure within the radius of its distance from Timber Lake and it is also one of the sections selected by Morgan and Petsch (1945, p. 26) to illustrate the work in which the Timber Lake received its name.

SECTION 8

PRINCIPAL REFERENCE, TIMBER LAKE MEMBER

Section 8 is almost entirely in the off-shore marine phase of the Timber Lake sand and illustrates well the rich faunas of the *Cucullaea* Assemblage Zone and adjacent levels. It's one principal lack as a representative of the southern aspect of the Timber Lake is the *Tancredia-Cymbophora-Pteria-Ostrea* beds at the top of the sand body; these are well exposed at the tops of nearby buttes to the north and northeast.

Exposures of the Timber Lake Member on the southwest side of a small butte on the divide between Red Earth and Meadow Creeks in NE ¼, SW ¼ Sec. 27, T. 15 N., R. 23 E., U.S.G.S. Lantry NE quadrangle, Dewey County, South Dakota (Loc. 34).

Thickness
(feet)

Top of Butte.

Fox Hills Formation (in part):
Timber Lake Member (in part):

12. Sandstone, light-gray and brownish-gray, thinly cross-bedded with dark-gray shale partings. A cap of indurated sand with few *Pteria linguae-formis*, *Ophiomorpha*, fragment palm frond; lies unconformably on under-lying unit as if slumped on it .... 5 to ?

11. Sand, fine-grained, weathers light brownish gray, chiefly massive, organi-cally worked, some clay streaks and blebs 22

10. Sand, fine-grained, highly glauconitic, with clay blebs and laminae; dark-greenish-gray, weathering to splotchy orange and brown stain with chips bright-orange ferruginous clay. Three layers punky, calcareous, glauconitic sand concretions approximately 3, 7 and 10 feet above base; lower con-tains abundant *Protocardia* 13.5

9. Sand, fine-grained, glauconitic, clayey; dark gray, weathering light gray with grayish-orange stain. A few scattered, punky, lime-cemented sand concretions, some with *Pteria linguaeformis* in upper two thirds. At base zone, similar concretions, *Pteria*, *Protocardia*, *Ostrea*, *Cymbophora* and *Cucullaea* (*Cymbophora-Tellina* Assemblage Zone) 19

8. Sand, very fine-grained, weathers yellowish gray; scattered punky concre-tions, some fossiliferous, *Pteria*, *Cucullaea*; in lower 2 to 3 feet scattered large concretions up to 2 feet diameter, sandy, punky, some with lime-stone core, fossiliferous: *Protocardia*, *Pteria*, *Cucullaea*, *Dentalium* 13

7. Sand, as above, scattered large *Cucullaea* and small concretions, some fos-siliferous: *Protocardia*, *Pteria*, *Sphenodiscus* 13.5

6. Sand, very fine-grained, silty, mottled with clay blebs; weathers yellowish gray; scattered limestone concretions in lower 5.5 feet, some with red-brown punky jackets, many fossiliferous; in upper 3.0 feet persistent zone large, red-brown-weathering limestone concretions, gray silty rinds; very fossiliferous *Cucullaea*, *Pteria*, *Protocardia*, *Ostrea*, *Sphenodiscus*, *Discosca-phites* (*Cucullaea* Assemblage Zone) 8

5. Silt, sandy and clayey, locally a silty sand, some dark-gray silty shale, weathers gray; 5 feet from top is persistent layer reddish-weathering bar-ren limestone concretions (barren B layer) 30

4. Silt, sandy as above, contains large punky concretions, gray silt rinds, fos-siliferous; and small hard blue-gray limestone concretions; *Oxytoma*, pro-tobranch bivalves. *Sphenodiscus*, *Discoscaphites*. (*Sphenodiscus* concretion layer with some *abyssinus* concretions) 3

3. Silt, sandy and clayey, weathers yellowish to brownish gray, scattered jarosite blebs and small, barren, reddish-weathering limestone concretions 5

2. Silt, clayey, some very fine-grained sand, weathers gray; scattered, chiefly barren, limestone concretions, one with *Actinosepia*; concretions of jaro-site nodules in basal 3 or 4 inches (upper jarosite layer) 13

Thickness Timber Lake Member measured .... 145 feet

Trail City Member (in part):

1. Silt, shaly, dark-gray, weathers light gray with yellowish cast; large red-brown-weathering, barren concretions with small cores and silt jackets at base (*Nucula* layer) 8

Cucullaea occurs throughout most of the interval between the *Cucullaea* Assemblage Zone and the *Cymbophora-Tellina* zone, and a few specimens are also found in the latter. The genera *Cymbophora* and *Tellina* are more abundant, at the level named for them, west of this section but *Cucullaea* continues to be present though rare and apparently only in concretions lacking *Cymbophora*.

SECTION 9

Section 9 is critical as the most complete one available on the Cheyenne-Moreau divide and the southernmost illustrating the marine-phase of the Timber Lake. Incidently, natural exposures of the Timber Lake sand-around this locality and the valley of Bear Creek just to the north are most likely the ones Hayden saw on his trip to Moreau trading post with Galpin and are as near as one can come to his "type" of Formation No. 5.

Composite section from adjacent exposures in dam spillway and cut of abandoned road (old U.S. 212) a little over a mile NE of Lantry in about center SW ¼, Sec. 4, T. 12 N., R. 22 E., Dewey County, South Dakota (Loc. 35).

	Thickness (feet)
Fox Hills Formation (in part):	
Timber Lake Member (in part):	
Grass roots top of road cut:	
6. Sand, fine- to medium-grained, greenish-gray, massive, weathers light yellowish brown, weakly indurated, friable with scattered *Cucullaea* in lower 12 to 15 feet. Scattered red-brown- to rusty-weathering calcareous concretions 0.5 to 1.5 feet diameter throughout, with hard sandy greenish-gray cores, some very fossiliferous, diverse fauna including *Pteria, Cucullaea, Oxytoma, Protocardia, Goniomya, Tellina, Sphenodiscus, Discoscaphites*	24
5. Sand, as above, with large rusty- to red-brown-weathering calcareous concretions up to 3.0 feet diameter in gray sandstone rinds; few scattered fossils	3
4. Sand, as in unit 6, with scattered *Cucullaea* and a few concretions, some with large *Sphenodiscus* and other elements of fauna as in unit 6	13.5
3. Sand, as in unit 6, with persistent large, rusty weathering calcareous concretions 2.0 feet or more diameter, some fossils, large *Cucullaea* common (*Cucullaea* Assemblage Zone)	3.0
2. Sand, very fine-grained; becoming clayey, scattered shaly partings; at base a layer thin, platey, barren limestone concretions 0.5 feet diameter (barren B layer)	3.5
1. Sand, very fine-grained, irregular lenses silty and clayey sand, shale blebs. Weathers gray to brownish gray; a few scattered *Cucullaea* and in lower 4 feet small calcareous concretions with *Discoscaphites nebrascensis* and *Cucullaea*	15
Thickness Timber Lake measured	62 feet

Water level in Spillway.

Identification of unit 3 as the *Cucullaea* Assemblage Zone in the presence of such an abundance of *Cucullaea*-bearing concretions is substantiated by the presence of the barren B layer of concretions below in base of unit 2, which, though inconspicuous, is widespread throughout the Moreau Valley area east of S. Dak. Route 65 (see unit 5 of Section 8 and fig. 25). The assemblage zone also crops out extensively just north of this section on either side of Bear Creek and its level can be related to the concretion layer, unit 3. This section records the maximum range of *Cucullaea* found in the area, approximately 55 feet, and indicates that the genus extends downward into the interval between the assemblage zone level and the *Sphenodiscus* concretion layer on the Cheyenne-Moreau divide.

SECTION 10

Section 10 includes the local Rock Creek lithofacies of the Timber Lake at its type locality, beds of the *Tancredia-Ophiomorpha* sand, and beds with the gradational *Dosiniopsis* fauna.

Composite section from exposures in bluffs on north side of the Grand River just west of the village of Bullhead in S ½, NW ¼, Sec. 24, T. 21 N., R. 24 E., U.S.G.S. Bullhead quadrangle, Corson County, South Dakota. (Loc. 25, 194).

	Thickness (feet)
Fox Hills Formation (in part):	
Timber Lake Member (in part):	
6. Sand, fine-grained, weakly indurated, friable; weathers massive, yellowish orange with few scattered limonitized shale partings and abundant, protruding, irregularly tabular, red-brown concretionary masses. A few scattered shells of *Tancredia*, rare *Ophiomorpha* in concretions	25
5. Sand, fine-grained, weakly indurated, weathers yellowish orange, contains scattered fossils and, at base, large, red-brown weathering, fossiliferous concretions up to 5 feet diameter. *Tancredia, Dosiniopsis, Ostrea, Lunatia* (= *Cucullaea* Assemblage Zone)	11
4. Sand, fine-grained, in thin tabular cross-laminated beds with some shale partings, gradational from unit below; at 9.0 from base layer flat-ovoid, barren calcareous concretions; at 8.0 from base a few scattered calcareous concretions with rare *Pteria, Ostrea*	16
3. Silt, silty shale and very fine-grained sand thinly interbedded and interlaminated, with sand increasing gradually upward; a few indurated lenses in upper half; local barren, platy, calcareous concretions 48 feet from base; gradual color change upward from gray to yellow gray with some orange-brown stain. (Rock Creek lithofacies, type section)	54
2. Silt, clayey, dark-gray, massive, weathers to light gray, 0.2 foot zone highly jarositic clayey silt at base (upper jarosite)	3
Thickness Timber Lake measured	109 feet
Trail City Member:	
1. Silt, clayey, dark-gray, weathers light gray, scattered jarosite stain, at base layer limestone concretions, red-brown-weathered rinds, some protobranchs (*Nucula* layer)	23

SECTION 11

Section 11 illustrates the contact with the Iron Lightning Member, the *Tancredia-Ophiomorpha* sand, and limitation of the sequence by change to the Irish Creek lithofacies.

Composite section from exposures in the valley walls of Hump Creek east of South Dakota Highway 65, in east half sections 30 and 31, T. 21 N., R. 23 E., Corson County, South Dakota (Loc. 48, 289).

	Thickness (feet)
Fox Hills Formation (in part):	
Iron Lightning Member, Bullhead lithofacies (in part):	
9. Shale, dark-gray, silt, and fine-grained sand thinly interbedded and lami-	

nated; silt and sand light gray with brownish tint, whole weathers a banded light brownish gray; local jarositic stain; becomes sandier upward   10 to ?

Timber Lake Member:

8. Sand, fine- to medium-grained, weakly indurated, dark-gray, slightly glauconitic, weathers yellowish orange; tabular, cross-laminated, few limonitized sandy shale partings; rare scattered calcareous sand concretions with *Ostrea, Pteria, Tellina, Oxytoma*   5

7. Sand, as in unit 8 above, in 3 to 8 inch tabular, cross-laminated beds, shot through with vertically and laterally extending, limonitized tube-fillings of *Ophiomorpha* and containing scattered shells of *Tancredia*   12

6. Shale, sandy, gray, with irregular, partly indurated masses of gray, glauconitic sand   2

5. Sand, as in unit 8, with *Ophiomorpha* locally abundant, some limonitized shale partings; at about 20 feet from base 0.5 foot bed of *Tancredia* shells, a few scattered in sand below this level   30

4. Sand, fine-grained, thin-bedded with some gray sandy shale laminae and partings; with rusty-brown-weathering concretionary masses of sandstone up to 2.0 feet diameter   4

3. Sand, fine-grained, massive, greenish-gray, weathers yellowish orange to yellow gray; scattered *Tancredia* shells in upper part. Lower 6 feet becoming clayey with some limonitized shale layers   20

Thickness Timber Lake Member   ....   73 feet

Trail City Member, Irish Creek lithofacies:

2. Sand, clayey, gray, massive, weathers light gray; in lower 2.0, scattered indurated gray sand lenses and 1.0 above base cone-in-cone masses in 2 to 4 inch green-gray bentonitic shale (D bentonite)   8

1. Sand, as in unit above, with scattered small indurated, locally fossiliferous sandstone concretions and, 13 feet from base, layer fossiliferous concretions 1.5 feet diameter; *Pteria, Cymbophora, Ostrea, Protocardia, Discoscaphites.* (=*Cucullaea* Assemblage Zone)   25

Bed of Hump Creek

SECTION 12

Section 12 is of a thin marginal part of the Timber Lake that illustrates faunal changes at the levels of the *Cucullaea* and *Cymbophora-Tellina* Assemblage Zones along the west side of the sand body.

Exposures at north end of high, west-facing cut of Standing Cloud Creek near its confluence with Firesteel Creek in SE ¼, NW ¼, SW ¼, NW ¼, Sec. 8, T. 18 N., R. 23 E., U.S.G.S. Black Horse S.E. quadrangle, Corson County, South Dakota. (Loc. 275)

Thickness
(feet)

Fox Hills Formation (in part):
Iron Lightning Member, Bullhead lithofacies (in part):

8. Shale, gray, thinly interbedded with silt and fine-grained sand, weathers light gray with brownish cast   7

7. Sand and shale, top and bottom foot massive, fine- to medium-grained sand, weathering yellowish green; middle part has thin interbeds fissile dark-gray shale   5

6. Shale, bentonitic, greenish-gray with thin layers greenish sand and scattered cone-in-cone concretions (E bentonite) 1

Timber Lake Member:

5. Sand, fine-grained, faintly cross-laminated, weathers light yellowish orange with abundant limonitic clay chips; abundant rusty *Ophiomorpha*, rare scattered *Tancredia* 15

4. Sand, fine-grained; tabular, cross-laminated, weathers light yellowish orange, abundant *Tancredia* and *Ophiomorpha* with bed of current-accumulated *Tancredia* shells concentrated in upper 4 inches........... 6

3. Sand, fine-grained, massive, weathers light yellowish orange; from 5 to 8 feet above base, rusty to red-brown irregularly tabular, sparsely fossiliferous calcareous sandy concretions with some *Tancredia*, few *Ophiomorpha*; associated small clusters *Ostrea* and *Pteria* (= *Cymbophora-Tellina* Assemblage Zone); in sand above scattered *Tancredia* and *Panopea* in living position 15

2. Sand, as in unit above, with some clay blebs, scattered small calcareous concretions in lower 3 feet, *Pteria* 3

Thickness Timber Lake .... 49 feet

Trail City Member, Irish Creek lithofacies:

1. Sand, clayey, fine-grained, gray with blebs gray clay; tough sandy gray calcareous concretions in lower 4 feet, fossils and abundant carbonized wood fragments in concretions and matrix; *Pteria, Cymbophora, Tellina, Tancredia* (= *Cucullaea* Assemblage Zone) 10

Creek level.

UPPER FOX HILLS FORMATION

MAJOR FEATURES

Two distinct but genetically related lithofacies make up the Upper Fox Hills Formation: 1) thinly interbedded clay, silt and very fine-grained sand that weathers to a markedly banded, light-brownish-gray outcrop, and 2) very fine- to medium-grained, clayey, subgraywacke sand that usually weathers to fluted light-gray or grayish-white outcrops but locally may be cemented to form resistant sandstone lenses. Along the Grand River and northeastward from the type area of the Fox Hills into the Missouri Valley area of North Dakota the sand lithofacies occupies the upper part and the thin bedded or "banded" lithofacies the lower part of the Upper Fox Hills sequence. It has been assumed that this simple relationship obtains throughout the Missouri Valley area of both North and South Dakota, although Laird and Mitchell (1942, p. 6) recognized that the sand, for which they used the existing name Colgate, was in some places partially or wholly replaced laterally by the banded bed lithofacies.

In the type area of the Fox Hills, where the two lithofacies have been formalized as the Bullhead and Colgate Members of the formation (Stevenson 1956, Curtiss 1952), the simple vertical succession does not hold south of the Grand River. In the Moreau valley area and beyond, the sand lithofacies recurs at more than one level in the westward-thickening Upper Fox Hills sequence and as many as 3 "Colgate" sands separated by banded beds may be present locally. The uppermost sand is commonly but not everywhere the thicker. One or all sands may be absent locally through lateral change to the banded bed lithofacies or, for the uppermost sand only, through removal by local channeling prior to the deposition of the overlying Hell Creek clays and lignitic clays.

Although it was in North Dakota that the term Colgate Member was first brought in and applied to the upper sand of the Fox Hills in the Missouri Valley region (Laird and Mitchell 1942), this member has never been employed there as a map unit. In the area of the type Fox Hills in South Dakota, Curtiss (1952) first used the name Colgate Sandstone Member in the Isabel Quadrangle but did not differentiate the unit on his map. Subsequent quadrangle maps of the South Dakota Geological Survey have shown the Colgate Member, but experience has demonstrated that it is not a useful map unit. It is discontinuous and highly variable in thickness, being generally under 40 feet and commonly less than 20. Were it not for the fact that it is topographically prominent at some places, forming the caprock of buttes where it has been locally indurated, it most likely never would have been used in mapping. Existing quadrangle maps show it as a continuous layer between the Bullhead Member and overlying Hell Creek Formation, yet in their descriptive text admit to its discontinuity. The false impression given is that a thin, hard, persistent, Colgate sandstone caps the Fox Hills Formation.

The concept of a widespread, generally indurated Colgate sand has led in many places to misidentification of the stratigraphic units represented in the numerous butte caps of the type area. This in turn has served to hide an interesting and perhaps geomorphically significant problem—the nature and origin of local siliceous induration in the Fox Hills, and some Hell Creek, sands. A thorough investigation of this "butte cap problem" is not within the scope of this study, but evidence pertinent to its solution, gathered during the field work, is given on pages 129-131. At this point the significant fact is that many butte caps that have been mapped as Colgate Member are not the upper sand; the outstanding examples are the locally indurated lenses in Timber Lake sand previously mentioned.

Colgate-like sand is common throughout the northern part of the western interior, occurring consistently in the sequence of beds transitional between the marine Fox Hills or equivalents and the continental Hell Creek (Lance) beds. In some places the white-weathering sands are stratigraphically concentrated, thick and persistent enough to be a useful map unit. This is not so in central South Dakota where thin lenses of the sand are commonly spread through much of the upper Fox Hills and lower hundred feet of the Hell Creek and only locally become the dominant lithology. Recognition of a Colgate Member here is more of a formality than a reality and on the grounds that it has little practical value and tends to obscure rather than elucidate the true stratigraphic relationships, the Colgate is dropped as a formal subdivision in the area of the type Fox Hills.

On the other hand, recognition of the lithofacies represented by the sands in question is important for environmental reconstruction. For this purpose the name Colgate lithofacies is applied here informally to any and all sand bodies similar in lithology to the upper sand. Likewise the Bullhead Member is reduced to the same informal status, becoming the Bullhead lithofacies. The underlying reason for this change in terminology is simply that these two distinctive lithologic types are so intimately associated and mixed that their separation is unnatural. Together they constitute a distinctive, easily differentiable, upper part of the Fox Hills Formation throughout the type area and adjacent outcrop areas in the Missouri Valley of North and South Dakota. The formal term Iron Lightning Member is applied here to this upper division of the Fox Hills Formation.

Iron Lightning Member

TYPE LOCALITY

The entire upper part of the Fox Hills Formation is exposed at a number of localities but one of the thicker, more informative and more easily accessible of these is an elongate badlands area about 3 miles southeast of the Indian village of Iron Lightning on the Moreau River. These exposures constitute the type locality for the Iron Lightning Member. They extend over 3 square miles in parts of sections 20, 28, 29, 32 and 33, T. 14 N., R. 19 E., and section 4, T. 13 N., R. 19 E., in Ziebach County. All parts of this badlands area except that in section 4 are shown on the U.S.G.S. Redelm NE Quadrangle. The area is approximately 5 miles north of U.S. Highway 212 on the graded road to Iron Lightning; the high western bluffs of the area lie a short distance east of the road.

A small area of the underlying Irish Creek lithofacies fortuitously exposed along a minor normal fault in the central part of the badlands makes it possible to see a complete section of the Iron Lightning Member within about ¼ mile radius of the center of the west half of section 33; the type section (#13, p. 132) is measured in this area. In addition, up to 80 feet of the overlying Hell Creek Formation is included, bringing the total thickness of beds in the exposure to about 250 feet. The type locality as a whole shows many of the lateral changes that take place within the member and illustrates especially well the variety of stratigraphic relations at the Hell Creek contact.

THICKNESS AND CONTACTS

The Iron Lightning Member varies considerably in thickness within the type area of the Fox Hills. The greatest thickness measured is approximately 170 feet (at the type locality), the least 58 feet. Lateral facies change of the lowermost part of the member eastward into the Timber Lake sand body accounts for at least some of the variation and for the consistently thinner sections overlying the sand body.

The basal contact of the member with both the Timber Lake and Irish Creek lithofacies is marked by a conspicuous change to the "banded bed" or Bullhead lithofacies. Generally at any one locality the top of the Timber Lake sand is a nearly plane surface on which lie the thin-bedded clays, silts and sands (Pl. 5). Sand beds just above the contact may be thicker and more numerous than usual for the Bullhead lithofacies and somewhat glauconitic like Timber Lake sand, weathering yellowish-gray rather than grayish white. At a few places the upper foot or more of the Timber Lake sand is thin-bedded and contains some shale partings. Contact with the Irish Creek lithofacies, seldom well exposed, is made evident by the change in outcrop color, from the dark gray of the latter to the light brownish-gray of the Bullhead lithofacies, as well as by the change upward to conspicuous thin-bedding. Locally, the upper several feet of the Irish Creek is thinly interbedded shale and silt or fine sand, the latter commonly impregnated with jarosite; more commonly, bedding is irregular or absent. Any silt beds at or close to the contact are likely to be jarositic.

The basal contact of the Iron Lightning Member is sharp and shows very few features in adjoining beds that could be classed as gradational. It is locally a time-transgressive contact, as is demonstrated by its abrupt rise in section eastward, within four miles along the Moreau Valley, from a position at the *Cucullaea* Assemblage Zone to one above the *Cymbophora-Tellina* Assemblage Zone. Relatively rapid local change in conditions of deposition accompanying a lateral shift in environment accounts for the abrupt change at the contact; no "break" in the sense of an appreciable hiatus is indicated.

The upper contact of the Iron Lightning Member, and of the Fox Hills Formation, is taken at the base of the first appreciably thick lignite or lignitic clay or shale. This old, arbitrary limit of the Fox Hills in its type area has proved a good operational solution to the problem of drawing a boundary within this variable sequence of beds representing the transition to continental deposits. It is expedient for the field mapper and useful to the environment-conscious biostratigrapher for in general it marks the beginning of coastal swamp and flood plain deposition.

At most places in the type area, sand of the Colgate lithofacies is overlain by dark brownish-gray or purplish-brown clay with a chowder of plant fragments and some scattered, thin streaks of vitrain. Strictly speaking, the basal bed of the Hell Creek is very rarely a lignite and seldom even a good lignitic shale, but it is a distinctive brown color that commonly weathers with a purplish cast; the content of fragmental plant remains is highly variable. Clay of this type is present at some places in the Iron

Lightning Member where it forms small lenses rarely more than an inch or two thick. The brown clay marking the base of the Hell Creek is variable in thickness but commonly is at least 3 feet thick and locally as much as 10 or 12 feet thick.

The "first appreciably thick" lignitic clay is not everywhere the same bed. The lower 50 to 100 feet of the Hell Creek typically contains a number of units of lignite and/or lignitic clay (see section 13) alternating with Colgate-like sands and gray bentonitic clay. The lignitic clays in the lower 10 or 20 feet of the Hell Creek are more lenticular and vary more in thickness laterally than those higher in the section. Even within small areas of 15 or 20 square miles it is not uncommon to find the "contact" lignitic clay thinning out locally. The next lignitic clay above then becomes the "contact" bed. Such local shifts in the contact are not generally great enough to be troublesome in mapping even at large fractional scales.

At some places the uppermost beds of the Iron Lightning Member have been channeled and the irregularities filled with either dark-gray Hell Creek clays or brownish-gray Bullhead-like clays (see Pl. 11). Because such channeling is local and the sediments in the channels contrast conspicuously in color and bedding structure with the Colgate lithofacies, it is relatively easy to discern. Channel-fills of Colgate-like sand may also occur but none have been identified; they would be almost impossible to distinguish from bodies of Colgate sand in the upper Iron Lightning Member unless exposed broadly enough to show their lateral relationships.

The Fox Hills—Hell Creek contact at one time gained considerable notoriety because of its involvement in the Cretaceous-Tertiary boundary problem. This was prior to 1930, and in the stratigraphic philosophy of the time, systemic boundaries were generally held to be unconformable. Attempts to show that the Fox Hills-Hell Creek boundary was unconformable were soundly disproved by Dobbin and Reeside (1929) in a special study of the contact in the northern Great Plains. Subsequently the transitional nature of the contact has not been seriously questioned. To characterize the change from Fox Hills to Hell Creek merely as a conformable contact is oversimplification. It was a complex change involving a variety of lithofacies deposited in marginal marine and brackish-water environments juxtaposed, along an irregular coast line, with others deposited in coastal swamp and related coastal plain environments. The time-transgressive lateral shift of one set of these environments over the other has been generally accepted for some time but few details of it have been worked out. Locally within the type area of the Fox Hills, relationships at the contact are complicated by interfingering of Hell Creek and Fox Hills lithofacies within a few tens of feet, and west of the type area there is broader interfingering on a much larger scale; both indicate fluctuations in the lateral shift of environments. These make difficult the consistent use of even an arbitrary contact and also serve to underline the obvious but frequently ignored fact that the genetic change from Fox Hills environments of deposition is not everywhere expressed in the stratigraphic sequence as a single, obvious lithologic contact.

BULLHEAD LITHOFACIES

The thinly interbedded sand, silt and clay of the banded bed, or Bullhead, lithofacies characteristically weathers to light brownish gray conspicuously banded badlands on steep slopes. Most of the beds range in thickness from a fraction of an inch to about 3 inches; some thicker beds of sand or clayey sand are scattered throughout the sequence. Prominence of banding is directly associated with degree of sorting, the more strikingly banded parts of the section showing the more complete separation into sand, silt and clay layers. Sand and silt relatively free of clay weather grayish or yellowish white, clayey sand and silt light yellowish gray to light brown, and clay gray brown, gray and dark gray. Subtle differences in value of gray mark slight changes in relative amounts of silt and clay. Flattened, ovoid, platy sand concretions with calcitic calcareous cement are common in the sandy layers; they generally weather a dusky rose color. Weathered outcrop surfaces are checked or, where bentonitic, have a "popcorn" crust.

Mechanical analyses made of samples of obviously more clayey parts of the Bullhead lithofacies in the type section of the Iron Lightning Member (Fig. 9) show that sand in the fine to very fine size range is the more common constituent in a majority of samples. Clay and silt together constitute from 25 to 50 percent of total weight in all 8 samples but were over 50 percent in only one sample and 30 percent or less in 5 sam-

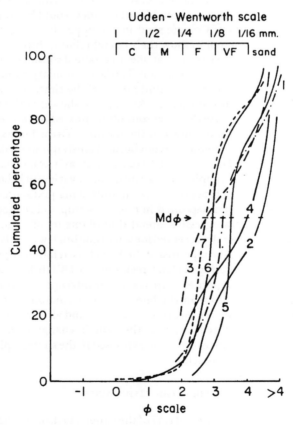

FIG. 9. Size composition of sands from the Iron Lightning Member (1-6) and lowermost Hell Creek Formation (7) in section 13 (loc. 74) represented by cumulative curves.
1-4; Bullhead lithofacies: 1) unit 3, 2) unit 6, 3) unit 8, five feet above base, 4) unit 8, 25 feet above base.
5-7; Colgate lithofacies: 5) unit 9, 6) unit 12, 5 feet from top, 7) unit 20 (Hell Creek), lower 5 feet.

ples. In the sand beds the sand is dominantly fine- and medium-grained and the silt-plus-clay fraction is generally less than 18 percent. Such bulk analyses of these variable beds are useful only in emphasizing that the Bullhead lithofacies, for all its superficial shaly appearance, is predominantly sand.

The conspicuous regular layering of the Bullhead lithofacies is apparent in the photographs of the Iron Lightning Member (Pl. 9, figs. A-B; Pl. 10, fig. C). Individual layers were not purposely traced in large exposures to see how far they continued, but several layers under an inch thick were followed for over 200 feet in one direction while piecing sections for measurement, and key beds—or groups of beds—several feet thick were matched in a series of sections extending over nine miles. On the other hand, local interruptions and discontinuities in the layering can be found at almost any exposure. Layers appear to terminate either through wedging out between thickening adjacent layers or by abrupt or gradual disappearance against erosion surfaces of shallow troughs. Rarely, layers are interrupted by local biogenic mixing of sediment, but this is exceptional. If the Bullhead lithofacies could be sampled by a multiple-frame coring device like those used in the study of recent sediments, the structure shown in the cores would undoubtedly be classed as a fine example of regular layering. Whether judged on details observable in such a limited example or seen over acres of outcrop, the thinness and horizontality of layering are the outstanding features of the bullhead lithofacies.

Internal structure of individual layers is somewhat variable. Sandy layers are commonly cross-laminated, some are massive and a few appear to be graded, but the latter are difficult to distinguish with certainty because of the relatively uniform grain size in individual layers and the obscurity of bedding planes between layers of only slightly different grain size. Spot examples of alternating coarser and finer layers can be found but no consistent sequential patterns were recognized. Massive clay layers with relatively little included silt or sand are not as common as other lithic types and are consistently thin, ranging in thickness from about `3 to 8 millimeters. Predominantly clayey layers thicker than this generally show fine mottling (Pl. 9, fig. C) which consists of pellets and variously bent, short, rod-like bodies of dark clay in lighter clayey silt or sand. Many clayey layers are obviously made up largely of these pellets, which appear compressed to varying degrees in the plane of bedding. These pellets, most likely fecal, are common throughout the Bullhead lithofacies and are one of its characteristic features.

Abundant plant fragments, another common feature of the Bullhead lithofacies, occur scattered throughout the layers and concentrated on bedding surfaces. In sandier layers they are locally abundant enough to form lignitic laminae. In general, they are too finely comminuted for identification, but rarely, coniferous twigs and reed-like monocots can be detected. Bits of charcoal are also common.

Biogenic structures other than fecal pellets are rare, but include local disruptions of layers and laminae. Irregular punctures filled with the matrix of an underlying layer can be found widely scattered throughout the lithofacies, but they are not common. Narrow, canoe-shaped troughs with concentrically laminated fillings in which plant and shell fragments are abundant (Pl. 9, fig. D) occur very locally associated with other invertebrate fossils at one of the few horizons at which the latter are common.

Sedimentary structures include local zones of spectacularly contorted bedding associated with intercalated lenses of sand of the Colgate lithofacies (Pl. 10, fig. A);

less commonly these occur independent of sand bodies. Intervals of contorted strata have variable but limited lateral extent and appear to be linear or elliptical in plan; outcrop measurements generally cannot be oriented as the contorted bodies are rarely exposed in more than one face. One contorted mass at Loc. 48 is at least 100 feet in long diameter, probably about 30 feet across and involves a sequence of beds at least 10 feet thick. At a different place (Loc. 93) a contorted zone involving a thickness of 15 feet of beds is only about 20 feet across. Edges of the contorted bodies, while fairly sharp and in some places channel-like, commonly show a tailing-off or shredding of layers into the contorted mass. Plastic deformation is abundantly evident in the contorted bodies but isolated, rotated blocks of "banded beds", some as much as 5 feet in long diameter, are also present locally. Contorted beds and laminae are continuous through concretions where a concretion layer is involved, indicating that the material later concreted was still plastic at the time of deformation (Pl. 10, fig. A).

Contorted strata show evidence of both hydroplastic deformation and block slumping. Where they form a sole layer to overlying Colgate-type sand bodies, as most of them do, load flowage is an obvious explanation. Either channeling and subsequent slumping took place locally along with the introduction of the sand, or hydroplastic flow was able locally to pluck out large blocks and carry them along without appreciably deforming them.

Evidence of other sedimentary structures is sparse. Because of the relatively unconsolidated nature of the beds, undisturbed bedding planes larger than hand-specimen size are rarely seen. Ripple marks were observed in very few places; all showed interference ripples and in one locality sand in the irregular trough pockets was semi-indurated with calcareous cement to form small light-reddish-brown concretions. Similar small concretions weather out of sandy layers at a number of places in most outcrop areas of the Bullhead lithofacies and presumably ripple marks are more common than is apparent.

COLGATE LITHOFACIES

The Colgate sand lithofacies of the Iron Lightning Member characteristically weathers to steep, white and grayish white, fluted outcrops on which scattered, rusty brown, calcareous concretions are locally conspicuous. At most places the larger bodies of sand form the upper part of the member; here they range in thickness from a few feet to over 60 feet. Sand bodies at other levels in the member are smaller and more sparsely scattered; these rarely exceed 20 feet in thickness.

Colgate sand in the type area of the Fox Hills Formation is a lithic sand which has been referred to in the quadrangle maps of the South Dakota Geological Survey as both a graywacke and a subgraywacke. If it seems helpful to use such terms, subgraywacke is the correct one. The sand is chiefly medium to very fine grained and, with some local exceptions, makes up 30 to 50 percent of the rock. Slightly altered, angular grains of quartz comprise half or slightly more of the sand fraction, and altered rock fragments commonly make up the remainder. The latter are chiefly quartzite, foliated rock fragments and chert; although partly sericitized these fragments have clearly delineated boundaries. A few glauconite grains are also generally present. The more common accessory minerals include biotite, chlorite, microcline, plagioclase, orthoclase, green hornblende, zircon, and leucite. Shell and plant fragments are present locally.

The clayey interstitial material, according to Ross (*in* Dobbin and Reeside, 1919, p. 25), is composed in large part of clay minerals that "appear to have formed in place by the alteration of detrital rock grains". The specimens examined by Ross were from Montana, but nothing was found that would contradict this observation in the limited examination of specimens from the type area of the Fox Hills. In the locally indurated lenses of Colgate the cement has yet to be identified; it is extremely fine and appears to be a mixture of siliceous and clayey material. Indurated lenses of both Timber Lake and lower Hell Creek sands have this same kind of cement.

The characteristic light gray to white color of the Colgate sands is attributed by Ross (*in* Dobbin and Reeside, 1929, p. 25) to "the white, chalklike appearance of the sericite and clay present as interstitial material between the grains . . ." The sericite rims of slightly altered quartz grains and sericitization of the siliceous rock types is also a contributing factor. Common dark minerals and rock fragments stand out sharply against the white background so that hand specimens appear peppered.

The Colgate lithofacies displays considerable variety in its modes of occurrence. Individual sand bodies may be relatively small and lenticular with an erosional lower contact, or they may be extensive and tabular with the lower contact surface non-erosional except for very local channeling. Commonly the basal part of a sand body is involved in the contorted structure mentioned in describing the Bullhead lithofacies. The dominant internal structure of sand bodies of the Colgate lithofacies is trough cross-bedding; this varies considerably in scale from place to place (Pl. 11, figs. A and C). Tabular cross-bedded structure is also fairly common. Bedding planes are accentuated locally by thin layers and laminae of carbonaceous matter and of clay, which is commonly iron-impregnated and weathers to bright-orange chips.

The appreciable content of plant debris in the Colgate sands gives rise to purplish-brown streaks and splotches on local outcrops. Loosely to tightly cemented, brown-weathering, limey, concretionary masses, roughly spherical to compressed ovoid in form, are a conspicuous feature of many Colgate sand bodies; their internal structure is identical to and continuous with that of the surrounding matrix. The diameter of the rounder concretions ranges from a foot to about 12 feet. At some places the more elongate masses reach 15 feet in long diameter and locally they coalesce to form larger bodies.

The lateral transition of Colgate sand bodies into the banded beds of the Bullhead lithofacies has two different aspects. In the larger sand bodies at the top of the Iron Lightning Member the internal structure first changes from trough cross-bedding to the tabular form with very low-angle, inclined bedding, or even to horizontal bedding. Intercalation of silty clay and shale layers between the sand beds follows. At this stage the individual sand beds, although typically Colgate-like in their overall characteristics, show much limonitic stain and jarositic impregnation and this feature together with the interbedded shale, which tends to be darker and grayer than that of the Bullhead lithofacies, completely changes the aspect of the sand body. At some places this conspicuously banded sand and shale continues to grade into the more subdued banding and brownish gray coloring typical of Bullhead lithofacies, at others it grades back into a Colgate sand body. The intermediate sand-shale phase does not seem to occur over appreciable areas—the greatest extent measured on the outcrop was about 300 yards. Lignitic shale and even thin seams of lignite commonly occur in it.

Smaller sand bodies that occur within the Bullhead lithofacies appear to develop locally at one or two fairly persistent horizons. For example, in the badlands of the type locality of the Iron Lightning Member, and in neighboring badlands within a radius of about 7 miles, one horizon at which lenses of Colgate lithofacies are locally developed can be identified in areas between sand bodies by the presence of scattered ovoid, friable, concretion-like masses of white-weathering sand (Pl. 10, figs. B-C) in a bed of sandy clay. These probably were formed from the foundering of small sand lenses in the clay, for their structural features are those of "pseudo-nodules" or ball-and-pillow structure. Where the interval becomes more sandy this type of structure gives way to sand-filled troughs and elongate sand lenses, which may or may not develop laterally into a conspicuous lens of sand.

Other sedimentary structures are as rare in the Colgate lithofacies as in the Bullhead lithofacies, although ripplemarks are commonly preserved where there is local induration of the sand.

FOSSILS

The Iron Lightning Member is not obviously fossiliferous but two characteristic modes of fossil occurrence in the Colgate lithofacies, 1) patch beds of *Crassostrea* and 2) channel-fills with abundant *Corbicula* and other animal and vegetable remains, account for scattered local accumulations of considerable abundance. In addition, sparse assemblages of fossils occur at some places in a few thin layers of the Bullhead lithofacies and the more resistant of these become scattered over the badlands slopes.

The fossil assemblages in the Bullhead lithofacies occur as scattered patches of fossil debris in thin beds of fine sand and on the associated bedding planes. Some of the shell material is fragmental but much of it is whole and well preserved, though commonly crushed. Clay fragments are common and associated thin clay layers may show pellet structure and sand-filled burrows. Locally the trough-shaped burrow fillings, previously noted (p. 121) are present. Although these separate biogenic structures may be abundant, thorough mixing of the layered sediment by organisms is uncommon.

Evidence in and around the type locality of the Iron Lightning Member in northern Ziebach County, where the member is extensively exposed, suggests that the fossil assemblages in the Bullhead lithofacies occur at a few preferred stratigraphic horizons where they persist, though discontinuously, over areas several townships in extent.

The position of fossiliferous horizons in several sections from this area is shown in Fig. 10. The most widespread horizon, that just above the second zone where lenses of Colgate lithofacies occur, has been traced over an area of approximately 100 square miles. Significantly, fossils in the Bullhead lithofacies are most common in the Moreau Valley area and to the southwest in western Ziebach and eastern Meade Counties. No comparable fossiliferous horizons were found either in the Bullhead lithofacies of the Grand River area or its continuation into the Missouri Valley area. This may be because outcrops are fewer and the Bullhead lithofacies commonly thinner, but enough of the northern exposures have been searched without success to keep alive the possibility that this distribution is a biogeographic feature, the incidence of marine communities increasing southward and southwestward in the Iron Lightning Member.

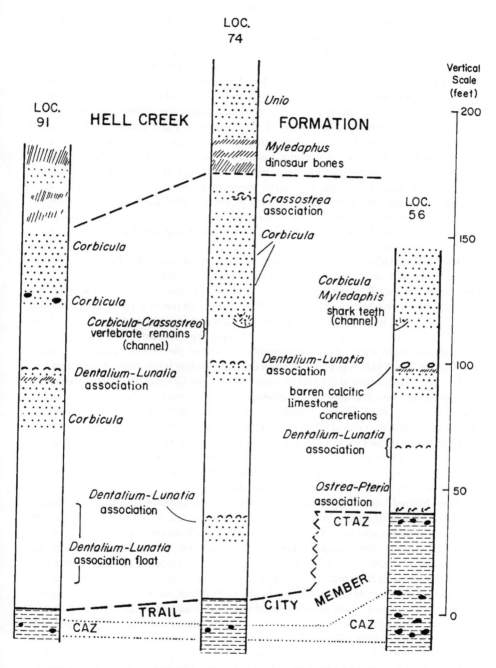

FIG. 10. Distribution of fossil assemblages in the Iron Lightning Member in the Moreau valley, Ziebach County. Colgate lithofacies indicated by stipple. Trail City Member is in the Irish Creek lithofacies which carries the *Cucullaea* Assemblage Zone (CAZ) and, locally, the *Cymbophora-Tellina* Assemblage Zone (CTAZ).

The small fauna of the Bullhead lithofacies is almost everywhere a *Dentalium-Lunatia* association. Hundreds of the small (1 to 1.5 cm in length) scaphopod *Dentalium (Antalis) pauperculum* Meek and Hayden occur with 4 other common molluscan species: the gastropods *Lunatia* (probably L. *concinna* Meek and Hayden) and *Piestochilus scarboroughi* (Meek and Hayden), and the bivalves *Spisula* sp. and *Nucula planimarginata* Meek and Hayden. *Lunatia* and the otolith *Vorhisia vulpes* Frizzell compete for the second most conspicuously abundant element of the fauna, but these weather out readily in good condition whereas specimens of *Spisula*, which are probably as common, are crushed and appear on the surface only as small fragments. The foregoing six species are the more common elements of the fauna which also includes small belemnite rostra, at least two species of *Discoscaphites*, *Cymbophora warrenana*, *Graphidula culbertsoni* (Meek and Hayden), an unidentified ostracod, worn fragments and small valves of *Crassostrea*, fish scales and vertebrae, shark teeth, and, more rarely, dental plates of a mollusc-eating fish *Mylognathus priscus* Leidy, scattered tooth elements of the ray-like *Myledaphus bipartitus* Cope, crocodile teeth and scutes, turtle scutes, water-worn wood and charcoal fragments and seed-like bodies.

The Colgate lithofacies, in its channel accumulations, its oyster-bed biofacies and in scattered shell layers throughout its sand bodies is characterized by the bivalve *Corbicula*. This easily recognizable little genus is a valuable indicator of brackish-water environments because it is restricted to sandy beds transitional between marine strata below and lignite-bearing continental strata above, throughout much of the interior Late Cretaceous in North America. Living representatives are found in both brackish and fresh water. Forms assignable to most of the several species originally described by Meek (1876, p. 520-524) from both the Judith River beds of Montana and the Fox Hills Formation of the Dakotas are found in the Colgate lithofacies of the Iron Lightning Member, Here specimens of several "species" with gradational forms occur in a single fossil assemblage and suggest that one polymorphic species populated the area of study. Until a critical study is made of the genus it is expedient to refer to all of the forms in the type Fox Hills as *Corbicula subelliptica* Meek and Hayden, the first specific name applied by them to specimens from the interior Cretaceous; their variety *C. subelliptica* var. *moreauensis* was based on specimens from the area that includes the type locality of the Iron Lightning Member.

The channel-bottom fossil accumulations of the Colgate lithofacies include a mixture of a few brackish- and fresh-water invertebrates, a considerable variety of vertebrate remains and some plants. Disassociated shell valves, bones, teeth, otoliths, scutes and wood are the common materials. *Corbicula* is by far the dominant fossil and well-preserved shells with both valves juxtaposed are commonly present. These two facts, together with the absence of the genus in the *Unio* association found locally in the channel deposits of the overlying lower Hell Creek beds and its absence from *Dentalium-Lunatia-Spisula* association of the Bullhead lithofacies, indicate that it was indigeneous to the environment in which the Colgate lithofacies was deposited, though not necessarily living at the particular spots where the channel-bottom accumulations are found. Of its associated fossils only the ray teeth *Myledaphus bipartitus* occur in enough abundance to suspect they too may have been indigenous. These teeth are also common in the overlying Hell Creek and similar, supposed fresh-water facies throughout the interior Cretaceous. Estes (1964, p. 160) suggested that the frequency of occurrence of *Myledaphus bipartitus* in the Lance Formation "may indicate that it was a permanent member of the fresh-water community". The abundance of its teeth in the channel bottoms of the Colgate lithofacies can be attributed to post-mortem accumulation at times of flood, particularly as it is rare in the oyster-bed biofacies and in the fauna of the Bullhead lithofacies. But its close association with great quantities of *Corbicula* suggest the latter as a possible food source; perhaps *Myledaphus* was equally at home in fresh and brackish water.

One of the most fossiliferous channel-bottom accumulations found is at the type locality of the Iron Lightning Formation (section 13, unit 12). The following list of fossils identified to date from this locality illustrates the nature of the channel accumulations.

Invertebrates

Crassostrea subtrigonalis (Evans and Shumard)
Anomia gryphorhynchus Meek
Corbicula subelliptica Meek and Hayden
Unio fragments
Viviparus cf. *retusus* Meek and Hayden
Goniobasis tenuicarinata Meek and Hayden
Melania insculpta Meek

Fish

Myledaphus bipartitus Cope
Vorhisia vulpes Frizzell (otolith)
Garpike scales
Unidentified fish bones

Reptiles

Turtle scutes
Crocodile bones and teeth
Hadrosaur teeth
Theropod claws
Dinosaur limb bones and vertebrae

Birds (bone fragments)

Mammals (teeth identified by James Hopson)

Cimolomys gracilis Marsh
Moeniscoessus robustus (Marsh)
Delphodon cf. *comptus* (Marsh)
Pediomys sp.

Plants

Conifer wood (trunks)
Conifer cones (probably *Sequoia dakotensis* Brown)

Oyster-bed accumulations of the Colgate lithofacies contain the most varied and abundant invertebrate fauna in the Iron Lightning Member. These occur chiefly as lenticular bodies of oyster-shell conglomerates with a sand matrix—mechanical accumulations, many of which are undoubtedly very close to the site of the oyster beds from which they were derived. The area covered by most of the oyster-bed patches is less than an acre and if they did not tend to form prominent ledges and caprocks, because of their relative resistance to erosion, they would seem rare. Stratigraphic distribution of the oyster beds is not completely known. Most occur in the Colgate sands at the top of the Iron Lightning Member, but a few are found on the south side of the Moreau-Grand divide that are very close to, if not continuous with, the top of the Timber Lake sand. Rare oyster accumulations also occur very locally above the lignitic clay that marks the base of the Hell Creek Formation.

The fauna from oyster beds at the top of the Iron Lightning Member is well illustrated by the accumulation that forms the caprock of Mud Butte, (Loc. 97) in the center W ½, sec. 25, T. 11 N., R. 18 E., Ziebach County, about 18 miles SSW of the Iron Lightning badlands. Stanton (1910, p. 179) lists 16 species from this locality (including 3 of *Corbicula* and 2 *Ostrea*). The following are the commoner species in our collections:

Crassostrea subtrigonalis (Evans and Shumard)
Corbicula subelliptica Meek and Hayden
Modiolus galpinianus (Evans and Shumard)
Anomia sp.
Melania cf. *insculpta* Meek
Melania wyomingensis Meek
Discoscaphites sp.

At Dog Butte, 6 miles east of Timber Lake in Dewey County (Loc. 16), an oyster conglomerate locally over 30 feet thick caps several small buttes and may rest directly on Timber Lake sand, although about 30 feet of grassed-over slope obscures the stratigraphy between Timber Lake outcrop and the butte cap. The following fossils occur in the Dog Butte assemblage:

Crassostrea subtrigonalis (Evans and Shumard)
Corbicula subelliptica Meek and Hayden
Anomia gryphorhynchus Meek
Tancredia americana Meek and Hayden
Indeterminate thick-shelled bivalve
Piestochilus scarboroughi Meek and Hayden
Melania wyomingensis Meek
Melania cf. *insculpta* Meek
Discoscaphites cheyennensis (Owen)
Discoscaphites nebrascensis (Owen)
Water-worn wood and bone fragments
Conifer cones
Shark teeth

The greater abundance of marine species in the oyster-bed associations and of fresh-water and terrestrial species in the channel-bottom association is apparent.

These distributions are summarized in the chart, Fig. 14. Together with faunas from adjacent parts of the lower Fox Hills and the Hell Creek Formation they constitute a remarkably clear record of the faunal transition from marine to continental environments.

The distinction between the channel deposits dominated by *Corbicula* and the oyster-bed deposits is largely based on the faunas. In a strict sense, either one may constitute a channel deposit and undoubtedly there are fossil accumulations that lie between the clear-cut examples presented here. The common occurrence of fossiliferous sand lenses containing abundant *Corbicula* and very little else would be an example. *Crassostrea* in great quantities, however, does appear to be restricted to patch-like lenses or to more tabular patch-like bodies; *Corbicula* lenses similar in form are on a smaller scale, feet rather than tens of feet. In addition, they are common in elongate sinuous bodies (true channel fills) where they are accompanied by vertebrate and fresh-water mollusc remains. Environmentally, the oyster-bed faunas probably indicate estuarine accumulations, dominantly brackish with closer marine than fresh-water associations, whereas the brackish *Corbicula* fauna has dominantly fresh-water associations and either represents deposits near the estuary head or in the distributary channels.

BUTTE CAP PROBLEM

Local induration of sand in the Fox Hills and lower Hell Creek formations has resulted in the formation of numerous buttes and ledges within the type area of the Fox Hills. Buttes with caprocks of indurated Hell Creek sand are few and the stratigraphic position of most can be demonstrated without much difficulty. The remaining butte caps in the area have collectively been considered to be indurated Colgate sand by most workers, some of whom have described this induration as a distinguishing characteristic of the upper part of the Colgate sand. The caprock of the buttes in question is a hard gray sandstone with siliceous cement. In color and in kind it is similar to fresh specimens of both the Colgate lithofacies and the sand in the upper part of the Timber Lake Member. Some caprock contains *Crassostrea* beds, some an *Ostrea-Pteria-Cymbophora* association, others a *Tancredia-Ophiomorpha* association, and still others are barren or contain plant remains. A majority of buttes have the slopes beneath the caprock obscured by grass and talus so that the underlying unit cannot be identified, but a few buttes with good exposures show that the caprock of some is underlain by the Bullhead lithofacies and of others by Timber Lake sand. Where the Bullhead lithofacies occurs the caprock is obviously indurated Colgate lithofacies. Where Timber Lake sand occurs the caprock has also been mapped as Colgate, and the Bullhead has been assumed to be missing from the section (Stevenson, 1960a).

During the present work on the type Fox Hills a number of exposures of indurated gray sand were found which are without question in the top of the Timber Lake sand. The better examples of these are along the Grand River west of South Dakota Highway 65, along Hump Creek east of the same highway, and in the less accessible valley of Firesteel Creek. On Hump Creek the Timber Lake sand is in the *Tancredia-Ophiomorpha* biofacies and generally soft and poorly exposed. South of Hump Creek in center NW ¼, NE ¼, sec. 31, T. 21 N., R. 23 E., a prominent, hard, gray sandstone ledge containing *Tancredia* and *Ophiomorpha* supports the east-facing edge of a bench, the top of which can be followed westward several hundred feet to a draw where typical, soft, dull-orange-weathering Timber Lake sand—also with *Tancredia* and *Ophiomorpha*—is exposed. It is apparent here that the sandstone is an indurated patch of sand identical to the Timber Lake; in the high bluffs a quarter of a mile to the south the Bullhead lithofacies comes in above the level of the bench and a thick Colgate sand succeeds it. Northeast of this ledge a half mile, across Hump Creek, a small butte is capped by a similar sandstone at the level of that in the bench, and upstream other ledges and small buttes can be found, all within the area of outcrop of the Timber Lake sand and below the Bullhead lithofacies.

Even more obviously within the Timber Lake sand are several similar ledges along the Grand River near the old village of Black Horse, from one to three miles west of S. Dak. Highway 65 bridge. Here local indurated bodies stand out prominently in the top of the Timber Lake and are directly overlain by the Bullhead lithofacies (Pl. 5, fig. C).

Buttes and indurated ledges are particularly abundant in the valley of Firesteel Creek throughout the 12 miles that it extends due south of its confluence with the valley of the Grand River. Here many small buttes, ridges and ledges are markedly elongate and directly associated with small normal faults. At several places the sandstone is in the form of a horizontal wedge, its broad end against the fault plane, the remainder tapering abruptly away from the fault. A good example can be seen in the southwest corner, sec. 8, T. 18 N., R. 23 E., where the west wall of the valley of Firesteel Creek transects a small east-west fault approximately at right angles. An elongate sandstone body abuts the fault on the north and, where exposed in the valley wall, is at least 15 feet thick near the fault but tapers out within 100 yards north of it. Three-quarters of a mile due east across the valley from this point the southeast face of a small butte, that may lie just north of the same fault, clearly shows the induration of the caprock cutting across the bedding of the sand so that 15 or 20 feet of hard sandstone on the southwest face is reduced to 2 or 3 feet on the northeast face some 100 feet distant.

Determining whether the Colgate or Timber Lake sands are involved in the indurated bodies along Firesteel Creek is complicated by the fact that the valley follows along the west side of the Timber Lake sand body where abrupt lateral changes to the Bullhead, and possibly Colgate, lithofacies take place. Although irrefutable proof of stratigraphic position is hard to find in the fault-bound sandstone bodies, at least two are in the top of the Timber Lake sand body.

The origin of the indurated lenses is a problem outside the context of this stratigraphic study, but the significant point to be made here is that their lithologic similarity is fortuitous and has no stratigraphic significance. This is indicated by 1) their spotty geographic distribution, 2) their occurrence at different stratigraphic levels ranging from the upper Timber Lake Member to the lower Hell Creek Formation, and 3) their occurrence in sand bodies that obviously formed in different environments of deposition. In the valley of Hump Creek and Grand River the distribution of indurated sand lenses on either side of the streams at one or more roughly concordant levels and, at some places, near the surface of broad benches, suggests their connection with geomorphic cycles although they lie well above recent terrace levels. A similar distribution of ledges is equally evident in the valley of Firesteel Creek but many indurated sand bodies in this area are unquestionably related to faults. In an appreciable number of examples from all the areas noted, the ledges are associated with both bench levels and faults, and conceivably a combination of geomorphic and structural factors is involved in their origin. These relationships, together with the three previously mentioned, point to local secondary —probably relatively recent—cementation; they are most certainly not a primary diagenetic feature restricted to one particular sedimentary unit. Although many ledges and butte caps are indurated parts of the Colgate lithofacies the induration is not characteristic of it and cannot be used as a clue to its identity.

On the Moreau-Grand divide east of the longitude of the valley of Firesteel Creek, buttes are numerous and their caprock is particularly hard to identify because most are erosional remnants on the surface of the divide and the indurated lenses cannot be observed in the context of surrounding sediments as they commonly can be in the major stream valleys. Here, too, the butte sides are usually covered so that one is confronted with a "floating" outcrop. No lithologic criteria are known that will serve to distinguish between a butte cap of Colgate sand in the Iron Lightning Member and one in the Timber Lake Member. Rock fragments in Timber Lake sand are more decomposed in general than those in Colgate sand but this distinction does not hold for sands of the uppermost part of the Timber Lake sand body—and this is the part of the Timber Lake that locally contains the indurated lenses. Moreover, some of the indurated sands at the top of the Timber Lake may indeed be Colgate lithofacies implaced in channel-like bodies in the Timber Lake sand. Paleontologic criteria may be helpful if not definitive. All *Crassostrea* accumulations whose stratigraphic position is known are in the Colgate lithofacies at the top of the Iron Lightning Member, but there are several of unknown position, such as that on Dog Butte, that lie very close to the Timber Lake sand body. Accumulations of the *Ostrea pellucida—Pteria linguaeformis* association have yet to be found in the main body of Colgate lithofacies at the top of the Iron Lightning Member and can probably be referred with reasonable confidence to the Timber Lake sand. Caprock with nothing but plant remains would not be likely in the Timber Lake, but it might be in Hell Creek. As is pointed out in the section on the relationship of different Timber Lake lithofacies, the lower Colgate sand bodies within the Iron Lightning locally appear to interfinger with or be emplaced in the Timber Lake sand. In such complex areas, assignment of butte caps to Iron Lightning or Timber Lake is completely arbitrary.

REFERENCE SECTIONS

The type section of the Iron Lightning Member (Section 13) is a composite of measurements made within an area of $\frac{1}{4}$ square mile in the central part of the badlands which constitute the type locality. Section 14, a reference section supplemental to the type, was measured about 4 miles due north of Section 13 in high bluffs on the north side of the Moreau River. It illustrates the local development of more than one prominent lens of the Colgate lithofacies. Section 15 illustrates a thinner sequence of the Iron Lightning Member farther east on the Moreau where the underlying Irish Creek lithofacies extends up to the *Cymbophora-Tellina* Assemblage Zone. In Section 16, on the Grand River, the Iron Lightning Member rests on the Timber Lake Member and the Colgate lithofacies predominates.

Section 13, type of the Iron Lightning Member, is also the principal reference section for its two lithofacies, both of which originated as formal stratigraphic units and hence have type exposures. The Colgate lithofacies needs a local principal reference section because the type Colgate Member is in Montana. The Bullhead lithofacies originated as a local member (Stevenson, 1956) but also requires a principal reference section because the type exposures do not show its thicker, fossiliferous phase and are not well enough exposed to provide a complete composite section.

SECTION 13
TYPE SECTION, IRON LIGHTNING MEMBER

Pieced from partial sections in W ½, Sec. 33, T. 14 N., R. 19 E., U.S.G.S. Redelm NE quadrangle, Ziebach County, South Dakota (Loc. 74). Units 1 thru 9 measured on exposures SE side of tributary gully, to west of main gully draining badlands, in SE ¼, NW ¼, Sec. 33. Units 10 thru 14 measured on SW corner of butte-like divide between the tributary and main gully in center SE ¼, NW ¼, SW ¼, Sec. 33. Units 15 thru 29 measured on NE-facing bluff of badland rim in about center of west line Sec. 33, just NW of low spur trending SE into badlands.

	Thickness (feet)
Clayey soil top of badland rim	
Hell Creek Formation (in part):	
29. Sand, fine- to medium-grained, subgraywacke, friable at base becoming clayey upward, some irregular sandstone masses with calcareous cement; jarosite masses and orange-brown ferruginous concretions in lower foot, ferruginous clay chips at top	13
28. Shale, fissile, carbonaceous, grading upward to purple-brown, lignitic shale	2
27. Silt, clayey, light-gray	2
26. Shale, lignitic, weathers purple gray	4
25. Silt, clayey to shaly, weathers light brownish gray, lenses fine-grained, laminated sand and silt; plant fragments; jarosite blebs and rusty, ferruginous concretions in upper 2 feet	9
24. Shale, lignitic, purple-brown, partings and lenses lignite and very fissile, silty shale	3
23. Shale, silty, brownish-gray, plant fragments, ferruginous concretions lower 2 feet	6
22. Lignitic shale, purple-brown...................................	2
21. Shale, finely silty, weathers greenish gray	3
20. Sandstone, fine- to very fine-grained, clayey, subgraywacke weathers to light gray fluted bluff; carbonaceous laminae reveal cross-lamination; some rusty laminae, ferruginous concretionary blebs and soft jarositic lumps, carbonized wood fragments common, few bone fragments and tooth; basal 5 feet with scattered pearly bivalved shells of unionid clams (lens of Colgate lithofacies)	23
19. Silt, shaly, light brown-gray, some lignitic shale layers	2.5
18. Shale, lignitic, and purple-gray clayey silt, scattered brittle ironstone concretions at top; reptile teeth and bone fragments..........	2.5
17. Silt, clayey, brown-gray	2

16. Shale, lignitic, lenses gray to brown-gray shale with plant fragments   6

Thickness Hell Creek measured   80   feet

Fox Hills Formation:
Iron Lightning Member:

Colgate lithofacies:

15. Shale, sandy to silty, gray to brown-gray, carbonized fragments, grades to silty clay in upper 3 feet, concentration of jarosite blebs at top   7

14. Sand, very fine-grained, clayey, weathers on slope to light gray crust, powdery beneath, probably bentonitic; carbonaceous laminae indicate cross-bedding   5

13. Silt, clayey, brownish-gray, hard, abundant carbonaceous fragments, some rusty stain   4

12. Sand, clayey, very fine- to medium-grained, subgraywacke; weathers light gray to grayish white, steep, fluted faces; thin layers orange ferruginous shale, lenses and layers *Corbicula* shells, some carbonaceous laminae, all show cross-bedding; huge brown-weathering concretions lime-cemented sand up to 12 feet long (usually vertical) diameter in lower $\frac{2}{3}$; base locally channeled, gradational in many places; fossils chiefly in lower 10 feet, chiefly *Corbicula;* local, abundantly fossiliferous basal channels *Corbicula, Crassostrea, Anomia,* coniferous wood and cones; abundant vertebrate remains, chiefly ray teeth (*Myledaphus bipartitus*), otoliths and dinosaur bones, but great variety of small fragments including mammal teeth   39

11. Sand, as in unit 12 above, mixed with irregular lenses silty gray shale, some carbonaceous silt, grades into unit above, is locally fluted   3

Bullhead lithofacies:

10. Shale, silty, fine-grained sand and clayey silt, thinly interbedded, weathers to striking banded outcrop with rock types dark gray, light gray and brownish gray respectively; basal 2 feet bentonitic, more shaly, locally weathers to gray, checked or "popcorn" crust; at base is fairly persistent fossiliferous horizon; abundant *Dentalium pauperculum, Lunatia,* shark teeth, otoliths, also *Piestochilus scarboroughi, Belemnitella, Nucula, Spisula, Discoscaphites*   18

Colgate lithofacies:

9. Sand, very fine-grained, clayey subgraywacke, weathers light gray, some thin interbeds gray to brown-gray silty shale, local conspicuous, gray-white ovoid, masses of cross-bedded fine-grained sand; (a persistent marker bed in the badland area)......................   5

Bullhead lithofacies:

8. Shale, finely silty, dark-gray to brownish-gray, locally bentonitic, thinly interbedded with light gray silt and in upper part some layers very fine sand; abundant fine plant fragments throughout, generally concentrated on bedding planes; whole weathers to crusty banded outcrop, banding not as contrasty as in units 10 above and 3, 4, 5 and 6 below; few scattered small reddish-gray, calcareous concretions   37

7. Shale, as in unit 8 above, thinly interbedded with very fine-grained, cross-laminated and cross-bedded sand; sand beds 1 to 1.5 feet

thick common at top and base; red-brown, cross-laminated calcareous sand concretions scattered and in local layers; unit locally more shaly ....   5⠀

6. Silt, clayey, silty shale and very fine-grained sand; thinly interbedded, conspicuously banded brownish gray, gray and light gray; beds up to 0.3 foot thick with some sand beds to 0.7 foot; clayey beds with abundant fine plant fragments and local patches with numerous small dark-gray blebs (probably fecal pellets); shell fragments and scattered otoliths and *Dentalium* chiefly in lower 4 feet   14

5. As in unit 6, with 0.9 foot sand layer at top containing large, irregularly flat-ovoid, calcareous sand concretions, weather red brown, blocky fracture (sand at top locally thickens to lenses of Colgate lithofacies)   8

4. As in unit 6, sand layers commonly have many flat-ovoid, red-brown calcareous sand concretions   13

3. As in unit 6 with small oval soft patches jarositic silt up to 4 feet from base, at top thin layer bentonitic shale capped by layer red-brown flat-ovoid concretions with cone-in-cone structure   5

2. Clay and shale, silty, dark-gray with irregular thin interbeds of very fine-grained light-gray sand; sand layers in lower 2 feet jarositic; whole weathers to checked and crusty gray "popcorn" surface with jarosite weathering out as yellow chips at base   5

Total thickness Iron Lightning Member   168   feet

Trail City Member, Irish Creek lithofacies:

1. Silt, clayey, gray, massive becoming shaly in upper few feet; few scattered fossiliferous limestone concretions about 12 feet from top *Cucullaea, Pteria, Protocardia, Discoscaphites,* (*Cucullaea* Assemblage Zone) ....   15 to ?

Gully bottom.

SECTION 14

High west-facing bluff dissected into badland topography east of the Moreau River in E ½, Sec. 9, T. 14 N., R. 19 E., U.S.G.S. Redelm NE quadrangle, Ziebach County, South Dakota (Loc. 91). Units 1 to 7 measured on SW face of spur in NW corner Sec. 9 and its short extension into Sec. 8. Units 8 to 19 measured on faces due west of triangulation station on high point of bluff in NE corner SW ¼, Sec. 9.

Thickness
(feet)

Top of bluff at triangulation station
Hell Creek Formation (in part):

19. Lignitic clay, ironstone concretions ....   2 to ?

18. Sand, very fine-grained, locally clayey and silty, local weakly indurated cross-laminated masses, weathers yellowish gray   17

17. Lignitic shale, seams lignite and interbeds gray shale; small ironstone concretions   9

16. Sand, very fine-grained, massive, weathers gray white, fluted, some lignitic shale partings near base   7

15. Clay and shale, silty, locally lignitic, weathers gray to purple gray; some small ironstone concretions   17

Fox Hills Formation:
Iron Lightning Member:
Colgate lithofacies:

14. Shale, dark-gray, interbedded with light-gray fine-grained sandstone in beds to 0.5 foot thick, sand decreases upward; gradational into unit 15 above 6

13. Sand, fine- to medium-grained, semi-indurated locally to cross-bedded and cross-laminated brownish-gray ledge; laterally to south changes to more typical fluted, light-gray badland slopes; local concretionary mass with *Corbicula* 25

Bullhead lithofacies:

12. Bentonite and bentonitic shale, greenish-gray to bluish-green, local sandy stringers, sandy at base; a very local accumulation in channel cut in unit below; cut out in turn by channeling at base of overlying sand .... 0 to 11

11. Sand, very fine-grained, interbedded with silty dark-gray shale, chiefly sand in upper 2 feet, scattered jarositic blebs 8

10. Silt, silty clay and shale, thinly interbedded and interlaminated, cross-laminated, weathers bluish to olive gray, weakly banded; at top persistent 0.8 foot sandy bentonitic shale weathers to dark crust; at base bentonitic shale up to 1.5 feet with local fossiliferous sand lenses in basal 0.5 foot; *Lunatia, Piestochilus scarboroughi, Crassostrea* (small), *Belemnitella*, otoliths, shark teeth, rare ray teeth (*Myledaphus*), crocodile teeth 15

Colgate lithofacies:

9. Sand, very fine-grained, clayey, subgraywacke, weathers to light gray fluted slopes; local gray-white ovoid sandy concretionary masses and some large rusty-brown calcareous sand concretions; locally a thin layer lignitic clayey sand present in upper foot 20

Bullhead lithofacies:

8. Shale, silty, gray, and brown-gray clayey silt thinly interbedded with light-gray silt and very fine-grained sand; conspicuously banded outcrop; upper foot is light gray bentonitic clay 20

7. Silt, clayey, and silty shale, thinly interbedded, with scattered thin layers light-gray silt, weathers light brownish gray, banding subdued; a bentonitic shale 11 to 12 feet from base, some red-brown calcareous silt concretions chiefly in layers 13 to 17 feet from base 38

6. As above, with more silt layers and persistent layer reddish-gray calcareous silt and sand concretions in very fine-grained sand bed at top 11

5. Covered by slope wash 3

Total thickness Iron Lightning Member .... 157 feet

Trail City Member, Irish Creek lithofacies:

4. Clay, silty, and clayey silt, dark-gray, with irregular thin lenses and pods of light gray jarositic silt; jarosite concentration at base 3

3. Clay, silty, bentonitic, weathers to dark "popcorn" crust (D bentonite) 1.5

2. Clay, silty, gray, massive, with irregular patches of silt; at base layer of widely spaced flat-ovoid calcareous concretions, cone-in-cone rinds, one with *Cucullaea* (*Cucullaea* Assemblage Zone) 2.5

1. Clay, silty, gray massive .... 10 to ?

Slope wash.

SECTION 15

Composite from exposures along ¼ mile of the southwest-facing badland scarp of St. Patrick's Butte (formerly Ragged Butte) between bench mark on high point of Butte in center E ½, SW ¼, SE ¼ and the SW corner, NE ¼, Sec. 23, T. 15 N., R. 21 E., U.S.G.S. Dupree NE quadrangle. Ziebach County, South Dakota (Loc. 56).

	Thickness (feet)

Grass roots near top of bluff.

Iron Lightning Member:

Colgate lithofacies:

21. Sand, fine- to very fine-grained subgraywacke, weathers light yellowish gray; with huge ovoid brown-gray-weathering calcareous sandstone concretions up to 6 feet long diameter .... — 7 to ?

20. Sand, fine- to very fine-grained, interbedded with thin layers and laminae silty to sandy, gray and dark-gray shale; abundant fine plant fragments; local small rusty-weathering ironstone concretions in sand layers .... — 4.5

19. Sand as in unit 21 above, with concretions .... — 2.5

18. Interbedded sand and shale as in unit 20 above .... — 2.5

17. Sand as in unit 21 above; with scattered small rusty concretions ... — 2.0

16. Interbedded sand and shale as in unit 20 above .... — 2.0

15. Sand as in unit 21 above, with concretions, weathers yellowish orange, cross-bedded .... — 4.5

Bullhead lithofacies:

14. Shale, silty, gray, clayey silt and yellow-gray to rusty-weathering fine-grained sand, thinly interbedded, lenticular; abundant plant fragments, some lignitic partings; basal 4.5 feet chiefly gray silty shale, rusty bands, remainder conspicuously banded .... — 13

13. Silt, clayey, grading in upper foot to a gray very fissile ("paper") shale; locally zone barren gray limestone concretions in lower foot . — 2

12. Silt and clayey silt with 0.4 zone lignite and lignitic clay in middle part .... — 1

11. Sand, very fine-grained, silty, locally clayey, laminated to cross-laminated weathers light gray, somewhat fluted; abundant plant fragments, flaky thin layers rusty ferruginous clay and silt; local calcareous, white-weathering, cross-laminated, friable ovoid bodies sand. Local masses with contorted bedding in lower parts of this and underlying unit (a thin Colgate lens) .... — 9

10. Silt, light gray, with some very fine grained sand, and gray shale, thinly interbedded, laminated and cross-laminated, abundant fine plant fragments on bedding surface, weathers light gray to light brownish gray; a few sandy silt beds as much 0.4 foot thick; silt layers with scattered small flat red-brown-weathering, calcareous silt and sand concretions .... — 10

9. Silt, and fine-grained sand, chiefly in platy, calcareous concretionary masses weathering white with some red stain .... — 1

8. Interbedded silt and shale as in unit 10 above .... — 6

7. Sand, very fine-grained, local semi-indurated gray-white calcareous masses and concretions; in beds about 1 foot thick interbedded with banded silt and shale as in units above and below .... — 6

6. Interbedded silt and shale as in unit 10 .... — 5

5. Clay, finely silty, gray, thinly interbedded and laminated with clayey silt, and very fine sand; may be bentonitic, covered with gray checked crust, persistent rusty-weathering layer near top 8

4. Interbedded silty clay and silt as in unit 5 above, clay waxy, commonly with bleb structure, probably bentonitic; whole weathers to gray "popcorn" crust; rusty- to yellow-weathering silty layers at base and 1.2 above base 15

Exposed thickness Iron Lightning Member .... 101 feet

Trail City Member, Irish Creek lithofacies:

3. Sand, very fine-grained, silty and clayey, massive, organically mixed, weathers light gray; locally with shells scattered or in subspherical clusters with *Ostrea*, chiefly in upper 2 feet; grades to unit 2 below, differs from unit 4 above chiefly in lack of sorting into thin alternating layers; *Lunatia concinna*, *Piestochilus scarboroughi*, *Graphidula culbertsoni*, *Rhombopsis newberryi* (Meek and Hayden), *Spironoma* sp., *Ostrea pellucida*, *Tellina*, *Nucula*, *Oxytoma*, *Pteria linguaeformis*, *Discoscaphites*, *Belemnitella*, otoliths, crocodile plates 4

2. Concretion layer, calcareous, with gray silt rinds, weather rusty brown, some fossiliferous; *Pteria linguaeformis*, *Cymbophora*, *Tellina*, etc. (*Cymbophora-Tellina* Assemblage Zone) 1

1. Silt, clayey and sandy, dark-gray, massive .... 5 to ?

Slope wash.

SECTION 16

Pieced from exposures on northwest-facing badland scarp on south wall of valley of Hump Creek in N ½, NW ¼, SE ¼, Sec. 31, T. 21 N., R. 23 E., U.S.G.S. Black Horse NE quadrangle, Corson County, South Dakota (Loc. 48).

Thickness (feet)

Top of scarp

Hell Creek Formation (in part):

12. Silt, light-gray, massive, locally indurated to platy gray-white ledge with some ferruginous stain; rare large pieces of wood ..... 3 to ?

11. Lignitic shale, lenses of lignite, weathers brown, commonly has gray blocky bentonitic shale at base 5

10. Sand, fine- to medium-grained, slightly clayey, cross-laminated, some lignitic laminae and small ovoid ferruginous concretions; weathers light gray, fluted 11

9. Shale, silty, light brownish-gray, abundant plant fragments; at top 0.5 foot lignitic sandy clay 3

8. Silt, light-gray, shaly, small scattered rusty concretions 4

7. Shale, locally lignitic, brownish-gray; from 2.5 to 4.0 from base silty shale with rusty stain; scattered rusty concretions in upper part and layer punky limonitic concretions at top 8

6. Shale, lignitic, silty, with pods and streaks vitrain; ironstone concretions with plant hash 0.5 to 1.0 from base, and second layer at top; scattered dinosaur bones, turtle shell, wood 2.5

Fox Hills Formation:

Iron Lightning Member:

Colgate lithofacies:

5. Sand, fine- to medium-grained, somewhat clayey subgraywacke; weathers to light-gray checked and fluted surface; lenticular cross-laminated structure revealed locally by lignitic and rusty laminae; about 25 to 35 feet from top huge spheroidal, brown-weathering calcareous sandstone concretions up to 10 feet in diameter; scattered small ferruginous concretions throughout 52

4. Sand, fine-grained, and silt, clayey, mixed, contorted beds and laminae; interbedded sand, silt and clay chiefly in lower part also contorted, including red-brown-weathering concretions with contorted laminae .... 10±

Bullhead lithofacies:

3. Shale, silty, gray to brownish-gray, thinly interbedded with light-gray silt and very fine-grained sand; whole weathers brownish gray, conspicuously banded; small flat rusty calcareous silt concretions in upper part; lower 4 feet chiefly shale with scattered sand layers 21

Total thickness Iron Lightning Member .... 83 feet

Timber Lake Member (in part):

2. Sand, fine- to medium-grained, subgraywacke, greenish-gray, slightly glauconitic, weathers light yellowish orange; tabular, cross-laminated lenses with local partings gray shale that weather to rusty chips; locally in upper 2 feet rare punky concretions with Pteria linguaeformis, Ostrea pellucida, Tellina sp., Oxytoma nebrascana, plant fragments 5

1. Sand, as above, lacks shale partings, penetrated by Ophiomorpha tubes, scattered Tancredia .... 8 to ?

Bottom of gully.

6. CORRELATION AND AGE

CEPHALOPOD DISTRIBUTION AND FAUNAL ZONES

Distinguishing useful faunal zones in the type Fox Hills and employing them in correlation is impeded by several factors: 1) the non-cephalopod molluscan fauna, in the present state of the old systematics, appears to be relatively uniform throughout the interior Campanian and Maestrichtian rocks; 2) marked stratigraphic differences in bivalve distribution within the type area are unquestionably local and ecology-controlled; 3) the type Fox Hills is part of what is probably the youngest marine Cretaceous terrain presently exposed in the interior and except for other exposures in the contiguous Missouri Valley outcrop area it has no known marine equivalents in the interior region.

The possibility exists that the entire type Fox Hills was deposited during too short a time for obvious genetic changes to take place in its faunas. The only group in which changes have been noted are the cephalopods. In the apparent absence or great scarcity of planktonic Foraminifera the cephalopods are the only group at all likely to furnish guide fossils in the present state of knowledge of the Fox Hills faunas. But if the distribution of bivalves is largely influenced by ecology, what of the cephalopods?

All of the non-ammonoid cephalopods except the nautiloid *Eutrephoceras* show some preferential distribution and even the latter appears to be largely restricted to the Assemblage Zones. The one belemnoid, *Belemnitella bulbosa* Meek and Hayden, has been found in all Assemblage Zones and in the Irish Creek lithofacies in the adult form. It is more common in the base of the Rock Creek lithofacies of the Timber Lake Member and in the fossiliferous zones of the Bullhead lithofacies but here only small, presumably juvenile, specimens occur. *Actinosepia* distribution, previously described (Waage, 1965), is restricted (Fig. 18).

Of the ammonoid cephalopods, baculites are exceedingly rare though possibly significant for general correlation. A small form similar to *Baculites columna* of the Gulf Cretaceous appears with the Gulf Coast scaphitid immigrants about the horizon of the *Limopsis-Gervillia* Assemblage Zone; only a few specimens have been found in the formation. *Sphenodiscus* is abundant in the Fox Hills but its distribution in the formation, described in part on p. 76, indicates strong ecologic control. Adults preferred areas nearer shore than those covered by the Little Eagle assemblage zones, which contain chiefly small forms and appear to have served as nurseries. No sphenodiscids have been found in the Lower *nicolleti* Assemblage Zone or in the upper Pierre Shale, though in Wyoming and Colorado they occur in sandy equivalents of the Mobridge Member.

139

Some of the more obvious distribution features of ammonoids in the type Fox Hills that suggest ecological control are: 1) the abrupt drop in section westward of large sphenodiscids within the type area; 2) the change in level of the distinctive *abyssinus* fauna characterized by many juveniles and some adults of *Discoscaphites abyssinus;* and 3) the great concentrations of *Scaphites* (*Hoploscaphites*) *nicolleti* in the Lower and Upper *nicolleti* Assemblage Zones. The accumulations of *S.* (*H.*) *nicolleti* may reflect the coincidence of mass killings with habitual periods of swarming, as at mating season, rather than some environmental control. On the other hand, the concentrations of these scaphitids with associated bottom-dwelling molluscs in community-like settlements could indicate that they too were largely benthic in habit and to some degree responsive to changes in the benthic environment. With more distributional studies of ammonoids it becomes increasingly apparent, as other workers are beginning to note (Ziegler, 1963), that these exemplary index fossils were not as insensitive to environment as has generally been assumed.

With this in mind, it is somewhat disconcerting to find that the only conspicuous change in the cephalopod faunas takes place across the change from the Trail City clayey silts to the Timber Lake sands. Large, fairly obese schaphitids similar to *Discoscaphites nebrascensis* (Pl. 7, fig. B), but lacking flank nodes except on the tip end of the body chamber, occur in all the Little Eagle zones. *D. nebrascensis* of the overlying Timber Lake is obviously a multinodose member of this lineage. Similar changes are found in *Discoscaphites cheyennensis* across the contact, somewhat more nodose forms occurring in the Timber Lake. *Discoscaphites abyssinus* and *D. mandanensis,* on the other hand, show no change across the contact in question. The difference between the Timber Lake and Trail City scaphitids may represent a valid evolutionary change with time, but the possibility cannot yet be eliminated that it is simply an ecological variation—greater nodation in the sandy facies—with only local time significance.

In my study of Fox Hills cephalopods to date only the morphologic changes in *S.* (*H.*) *nicolleti* appear to indicate a reliable evolutionary trend, and it too undergoes one change about where the Trail City grades to the Timber Lake. *S.* (*H.*) *nicolleti* occurs in both the Trail City and Timber Lake Members, and a variant also occurs in the Moreau bridge faunule within the Mobridge Member of the Pierre Shale (p. 51). *S.* (*H.*) *nicolleti* has a fairly complicated rib pattern on its exposed whorls (Fig. 11), culminating in a zone of very fine ribs on part or most of the body chamber. As Figure 11 shows, the Trail City *nicolleti* has between half and two-thirds of its body chamber covered with fine ribs; the Timber Lake *nicolleti* has only the last third of the body chamber covered by the fine ribs. Examination of the *nicolleti*-like scaphitids from the Moreau bridge faunule shows that the entire body chamber is covered by fine ribs. In other words, there is a progressive forward shift of the rib pattern on *nicolleti* conchs at three different levels in the type area. In the hundreds of *nicolleti* specimens examined from the Trail City none of the Timber Lake-type rib patterns was found; only a single somewhat dwarfed specimen of the Mobridge type was found in a Lower *nicolleti* concretion.

Nicolleti-like scaphitids are one of the more common elements in the indigenous Late Campanian-Maestrichtian scaphitid faunas of the interior Cretaceous. Recognition of the shifting pattern of ornamentation may prove valuable in relative placement of isolated faunas like that of the type Fox Hills. To date it is the only feature of the cephalopod fauna of the type Fox Hills that seems to offer a reliable means of zonation within the formation, as well as relating it to other Fox Hills terrains.

The change in *nicolleti* within the Fox Hills lends some validity to the other slight changes noted for some of the ammonites. If these changes are indeed successional, they are relatively slight; the cephalopod fauna as a whole remains much the same —except for its peculiar distribution features—throughout the formation.

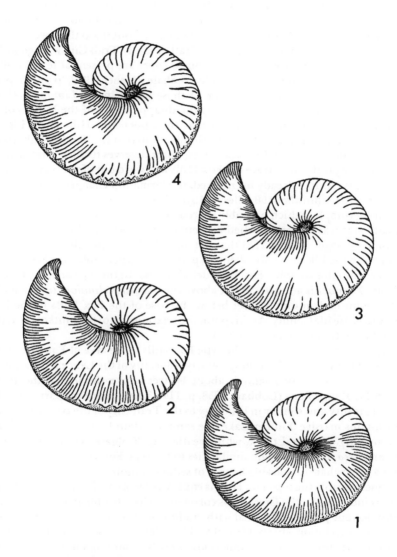

FIG. 11. Distribution of fine ribbing on shells of successively younger (ascending) variants of *Scaphites (Hoploscaphites) nicolleti.*(Diagrammatic sketches approximately ⅘ natural size)
1) Mobridge Member, Pierre Shale, Dewey Co., South Dakota;
2) Fox Hills Formation, Lance Creek area, Wyoming;
3) Trail City Member, Fox Hills Formation, type area;
4) Timber Lake Member, Fox Hills Formation, type area.

PROVINCIAL CORRELATION

The Fox Hills Formation can be traced northeastward from the type area into North Dakota. Here the most recent work on it in the Missouri Valley area of the state is included in the geologic study of Emmons County (Fisher, 1952), the southernmost county east of the Missouri River. Although the formation is changing in this direction the outcrops near the river clearly show the Timber Lake and Iron Lightning Members. Fisher recognized a "Trail City", but it was the lower clayey part of the Timber Lake Member. A few feet of rather sandy Trail City with good Lower *nicolleti* and *Limopsis - Gervillia* Assemblage Zones are present at least locally and usually have been included in the Pierre Shale (Fisher 1952, p. 9); reports of *Sphenodiscus* in the "Pierre Shale" are of those found in the *Limopsis - Gervillia* Assemblage Zone. Fisher (p. 12) reports the disappearance of the lower Fox Hills eastward within the county but notes that the upper Fox Hills (Iron Lightning Member) continues eastward. This may simply mean that the Timber Lake has gone laterally into an Irish Creek-like lithofacies, but it deserves further investigation.

It is far more difficult to trace the Fox Hills westward out of the type locality into Meade County. As has already been noted, the formation changes facies completely southwestward within about 50 miles (p. 15) and becomes a dominantly sandy, locally coal-bearing complex which probably represents a fairly persistent deltaic area. This enigmatic area was recently studied by Pettyjohn (1964) who documents an earlier conclusion (Waage, 1961, p. 238) that not only the beds equivalent to the type Fox Hills but also the Elk Butte Member and part of the upper Mobridge Member of the Pierre pass into this complex Fox Hills deltaic phase. In the type area the base of the Trail City Member is at least 250 feet above the *Baculites clinolobatus* Range Zone, but Pettyjohn finds it only 20 feet below the base of the Fox Hills in the Enning Quadrangle, Meade County. At this point the relationship of the units of the type Fox Hills to the delta complex is unknown.

Of special interest relative to the type Fox Hills is the area of Fox Hills outcrop northeast of Lance Creek, Wyoming—the type locality of the Lance Formation. Here the base of the Fox Hills sand is about 60 feet above the top of the *Baculites clinolobatus* Range Zone (Cobban, 1958, p. 114). The uppermost part of the Pierre Shale here is silty and similar in lithology to the Trail City of the type area. The Fox Hills contains massive Timber Lake-like sands overlain by more thin-bedded sand and commonly by several thick, white-weathering "Colgate" sands with *Corbicula* and *Crassostrea*. The lithologic similarities to the type Fox Hills are quite local but suggest a somewhat comparable history of sedimentation.

Unquestionably the type Fox Hills is equivalent to a considerable part of the type Lance, as indicated in the Cretaceous correlation chart (Cobban and Reeside, 1952, p. 1025). Recent papers concerned with eradicating the Danian from the American Cretaceous correlation charts (Jeletzky, 1962; Jeletzky and Clemens, 1965) have mentioned the discovery of the innermost whorls of an ammonite, identified as a scaphitid, from nonmarine beds high in the type Lance. But there is more substantial evidence for the correlation than this. The type Fox Hills contains both Lance mammals and Lance plants (p. 127). In addition, the Fox Hills of the Lance Creek area contains a variant of *Scaphites (Hoploscaphites) nicolleti* in which the fine ribbing covers almost the entire body chamber—to within a few millimeters of the ultimate septum (Fig. 11). In other words, the ribbing pattern of this form is intermediate between the *nicolleti* from the Mobridge Member of the Pierre Shale and that from the Trail City Member of the Fox Hills. This is good substantiation of the correlation of the Elk Butte Member of the Missouri Valley Pierre with the Fox Hills of the Lance Creek area—a correlation suggested orally to me by William A. Cobban many years ago.

AGE

Specialists agree that the fauna of the type Fox Hills Formation is Maestrichtian in age but there is uncertainty as to precisely what parts, or part, of the Maestrichtian are represented. Jeletzky (1960, 1962) offers substantial arguments for the Early Maestrichtian age of at least the lower part (Trail City-Timber Lake) of the type Fox Hills, but I do not find his evidence completely convincing. A review of the problems inherent in the long-range correlation of the type Fox Hills will, I believe, illustrate why a more definitive assignment than Maestrichtian seems premature.

Foraminifera and belemnoids are the principal zonal guides in the Eurasian Late Cretaceous, the use of both being enhanced by modern monographic studies and analyses. Neither the Fox Hills Formation nor the underlying uppermost part of the Pierre Shale preserve planktonic or other diagnostic Foraminifera in the type area and *Belemnitella bulbosa*, the only belemnoid in these beds, is apparently of little use. According to Jeletzky (1960, p. 31) both *B. bulbosa* and a closely related species from Selma and Ripley beds of the Mississippi embayment area "are immediate descendants of the Late Campanian *Belemnitella americana* (Morton, s. lato) of the northern Atlantic States and only superficially similar to *Bel. junior* Nowak s. str. of northern Eurasia." In other words, *B. bulbosa* of the interior is not diagnostic below stage level; it has no close relative in the Eurasian Late Cretaceous belemnoid sequence and being an end member of the last-surviving North American belemnoid lineage the uppermost age limit of its range can only be determined from associated fossils.

The ammonoids *Sphenodiscus* and *Scaphites* and the bivalve *Inoceramus fibrosus* have been the principal fossils used in correlating the type Fox Hills with the European Maestrichtian. Jeletsky (1962, p. 1011-1014) bases much of his argument for the Early Maestrichtian age of the type Fox Hills on *Inoceramus fibrosus*. On the basis of external morphology alone, he synonomizes under it the Eurasian Late Campanian-Early Maestrichtian form *I. caucasicus* Dobrov, and considers that all rocks with *I. fibrosus* in the western interior are, therefore, "equivalent to the late upper Campanian and lower Maestrichtian of northern Eurasia . . ." (p. 1013). Speden (1965, unpublished) points out in his redescription and reassignment of *I. fibrosus*, that similarity of hinge and musculation must be proved to establish the identity of *fibrosus* and *caucasicus*; and further, even if their identity is established there is no reason to expect the ranges to be identical in Eurasia and North America. Jeletzky (1962, p. 1013) notes also that in Russia *I. fibrosus* (=*caucasicus*) appears to be replaced by *I. tegulatus* Hagenow, in the Late Maestrichtian. In the western interior *I. fibrosus* ranges as high in the Cretaceous section (Timber Lake Member) as off-shore marine faunas are found. Just because *tegulatus* does not succeed *fibrosus* in North America one cannot assume that the latter did not exist there into the Late Maestrichtian. Until the taxonomy and distributional features of *I. fibrosus* are better understood, its uppermost age limit in the interior of North America, like that of *Belemnitella bulbosa,* can only be determined by its relation to more precise zonal indices.

The genus *Sphenodiscus* presents provincial problems in correlation and until the succession and biogeography of its species are worked out satisfactorily in North America it can hardly be expected to serve as more than a general index in correlation with the European Maestrichtian. The following succession of sphenodiscids from the Escondido Formation in northern Mexico, presented by Böse and Cavins (1927, p. 50) and apparently still accepted and relied upon (Young, 1960, p. 256; Howarth, 1965, p. 403), is the most complete sequence recognized in North America (descending order):

A. Zone of *Sphenodiscus pleurisepta* (Conrad)
B. Zone of *Coahuilites cavinsi* Böse
C. Zone of *Sphenodiscus intermedius* Böse
D. Zone of *Sphenodiscus lenticularis* (Owen)
E. Zone of *Coahuilites sheltoni* Böse

The distribution of these zones relative to another is questionable. From the description of the field occurrences (Böse and Cavins 1927, p. 45-56; Böse 1927, p. 184-188) it is obvious that no two zones were observed together at the same locality, nor was any formal attempt, such as measuring sections, made to match the stratigraphy from one locality to the next. The three upper zones occur in different areas and succeed each other geographically northward toward the Rio Grande.

From the Fox Hills Formation in the Lance Creek area of Wyoming I have collected four kinds of sphenodiscids in concretions scattered within 15 or 20 feet of vertical section. One relatively smooth species with a sharp venter is similar to the species in the type Fox Hills and can be referred to *Sphenodiscus lenticularis*. A second, ornamented with nodes on the flanks and with a broadly rounded venter, is a *Coahuilites* similar to *C. cavinsi*. A third is ornamented but has a sharp venter throughout and is probably a variant of *Sphenodiscus pleurisepta,* which it closely resembles in ornament and suture. A fourth is an undescribed form with ornament like *S. pleurisepta* and a rounded venter internally that becomes sharp on the outer whorl and body chamber. This assortment of sphenodiscids from one bed in the interior Cretaceous includes forms very close to, if not specifically identical with, three of the five zonal indices described by Böse and Cavins (1927). Taken together with the lack of stratigraphic control in Böse and Cavins' zonal arrangement they cast serious doubt on the reliability of these zones.

The smooth sphenodiscids, which are generally classed as *S. lenticularis* if found in the interior, or as *S. lobatus* (Tuomey) or *S. tirensis* Stephenson if found on the Atlantic or Gulf Coastal Plain, appear to range throughout the entire marine Maestrichtian in North America. In northern Europe smooth sphenodiscids are apparently the only kind present and these are restricted to the Late Maestrichtian, although as Jeletzky (1960, p. 32) points out they occur in the Early Maestrichtian in the Aquitanian basin and probably migrated northward only late in the stage. The great individual variation in suture within this group of smooth *Sphenodiscus* does noţ promise clear-cut specific distinction that might be helpful in a refined correlation of Maestrichtian zones. The known predeliction of sphenodiscids for marginal environments, so well illustrated in the type Fox Hills, is perhaps the strongest indication of all that the group cannot be expected to furnish reliable zonal indices for detailed biostratigraphy.

Scaphitids of the type Fox Hills, with the exception of some of the Gulf Coast elements previously noted, appear to be endemic species. The only species among them for which more than provincial distribution has been claimed is *Scaphites* (*Hoploscaphites*) *nicolleti* (Owen). At the subgeneric level Birkelund (1965) has shown that a close relationship exists between the *Scaphites* (*Discoscaphites*) of the type Fox Hills and those of the Maestrichtian of West Greenland; but comparable scaphitids are not certainly known from the European Maestrichtian so the group of forms currently classed as *Discoscaphites* is of no aid in zonal correlation. On the other hand *Scaphites* (*Hoploscaphites*), the prevalent scaphitid group in European Maestrichtian, is abundantly represented in the western interior of North America. Unfortunately the taxonomy of this group is not adequately understood either in North America or in Europe, where a considerable variety of forms are grouped under *S.* (*Hoploscaphites*) *constrictus* Sowerby. Scaphitids very similar to more than one of the variants of *S.* (*H.*) *constrictus,* including forms very close to Sowerby's type *constrictus* from Normandy, occur in the Fox Hills Formation in the Lance Creek area of Wyoming. No *constrictus*-like species, unless one accepts *S.* (*H.*) *nicolleti* as *constrictus*-like, occur in the type Fox Hills.

Jeletzky (1962, p. 1014-1016) is confident that a specimen he found in the lower Maestrichtian of Hemmoor, Germany, is conspecific with *S.* (*H.*) *nicolleti* from the type Fox Hills. I agree with Birkelund (1965, p. 158) that the Hemmoor specimen is not well enough preserved for certain identification, and I would add that what features it does show do not correspond with the distinctive pattern of ribbing on the Fox Hills specimens of *nicolleti.* Inspection of the adjacent figures 2A and 1B in Jeletzky (1962, pl. 141), which show, respectively and in comparable orientation, the Hemmoor specimen and a *nicolleti* from the type Fox Hills, reveal a different pattern of ribbing on the adapical parts of the hooks in the two specimens. Instead of the *nicolleti*-like bunching of fine ribs near the umbilicus, the Hemmoor specimen shows prominent, more widely and evenly spaced primary ribs. Jeletzky (1962, p. 1015) also synonomizes *S.* (*H.*) *constrictus* var. *tenuistriatus* Nowak under *nicolleti* on the basis that Nowak's figures of nodose *tenuistriatus* (Nowak, 1911, pl. 33, figs. 11, 12, 14 and 22) appear "indistinguishable" from the Hemmoor specimen. Nowak's *tenuistriatus* does appear *nicolleti*-like and although his figures are not adequate basis for comparison they superficially resemble the Mobridge *nicolleti* (p. 140) in the extent of the fine ribbing and the presence of both nodose and nodeless forms more than they do Fox Hills *nicolleti.* The presence of *S.* (*H.*) *nicolleti* in the European Early Maestrichtian has yet to be convincingly demonstrated, but even the identification of the species at this stratigraphic level would not in itself mean that the type Fox Hills was Early Maestrichtian, for the stratigraphic range of the lineage as a whole has yet to be worked out relative to the European stages; *nicolleti* such as those in the Mobridge Member of the Pierre Shale are most likely Early Maestrichtian, and the Trail City and Timber Lake *nicolleti* could easily be Late Maestrichtian.

None of the specimens of *S. (H.) nicolleti* examined from any of the four known levels of this lineage (Fig. 11) show umbilical bullae, a feature that Jeletzky accepts as characteristic of what he calls *S. (H.) aff. nicolleti* from the *Baculites grandis* zone. This latter type of scaphitid, a specimen of which was figured by Elias (1933, pl. 39, fig. 3) as *Discoscaphites abyssinus* (Morton) and more clearly refigured by Jeletzky (1962, pl. 141, figs. 3a, b, c.) as *S. (H.) aff. nicolleti* (Morton), is common in the Fox Hills Formation at Lance Creek, Wyoming. Here it occurs in concretions with characteristic *S. (H.) nicolleti,* from which it differs not only in the possession of umbilical bullae but also in its smaller umbilicus, generally smaller size and wider distribution of ventrolateral nodes on both the body chamber and the exposed phragmocone. This bullate form, which resembles Sowerby's type *S. constrictus* and other specimens of this species from Normandy, is distinct from members of the *nicolleti* lineage and it is an overgeneralization to group the two together as Jeletzky does (1962, p. 1015). In addition to the morphologic differences noted, the ranges of these two types of *Hoploscaphites* are not coextensive in North America. The *constrictus*-like forms have not been found with the upper two variants of *nicolleti* in the type Fox Hills and have not been reported in beds older than those of the *Baculites grandis* zone.

The *S. (H.) nicolleti* lineage undoubtedly has its roots in the variable species *S. (Hoploscaphites) gilli* Cobban and Jeletzky (1965) recently described from western interior strata where it ranges from the zone of *Baculites perplexus* Cobban to that of *Didymoceras stevensoni* (Whitfield). Thus the *nicolleti* lineage is distinct in the western interior at least as far back as the middle of the Late Campanian. As Cobban and Jeletzky (1965, p. 799) state, *S. (H.) gilli* seems most closely similar to the European *S. roemeri* d'Orbigny as interpreted by Schlüter (1871-72, p. 89; 1876, p. 163). Birkelund (1965, p. 110) synonymizes some of the forms so interpreted by Schlüter under *S. (H.) greenlandicus* Donovan. Apparently the *gilli-greenlandicus*-like *Hoploscaphites* were common to the western interior of North America, Greenland and northern Europe during at least part of the Upper Campanian. *S. (H.) gilli* appears to be smaller in average size than the later *S. (H.) nicolleti* but the shape and proportions of the shell are the same; some *gilli* have ventrolateral nodes, more lack nodes, but none have umbilical bullae.

The *gilli-greenlandicus-nicolleti* kinds of *Hoploscaphites* appear to form a natural group which is distinct from other somewhat similar *Hoploscaphites* such as the North American forms called *S. (H.) aff. nicolleti* by Jeletzky (1962, p. 1015) and similar European forms included in the loosely defined *S. (H.) constrictus.* At present *Hoploscaphites* seems to be the only group of fossils likely to permit a refined zonal correlation of the western interior Maestrichtian with that of Eurasia, but the taxonomy and biogeography of the group must be much more thoroughly understood before it can be used effectively for this purpose.

7. ENVIRONMENTAL HISTORY

SUMMARY OF EVENTS

Change from the broad lithologic homogeneity of the underlying Pierre Shale to the more locally diversified lithology of the Fox Hills is indicative of the change from more uniform offshore environments to marginal marine environments that become progressively more varied as the marine-nonmarine transition progresses. The widespread, finely silty shale of the Elk Butte Member of the Pierre was deposited where bottom conditions were either inhospitable to organisms or to their preservation; that it was even more restrictive than the environments of most of the underlying Pierre Shale is suggested by the scant Elk Butte fauna of arenaceous Foraminifera and linguloid brachiopods. This combination is usually taken as indicative of relatively shallow, brackish water; whether such conditions prevailed during Elk Butte deposition, about which virtually nothing is known, remains to be substantiated. Considering the rich marine molluscan faunas that subsequently inhabited the area, brackish conditions seem unlikely.

The influx of silt that marks the beginning of Fox Hills deposition indicates an increase of current energy in the environment and the lower Trail City beds provide evidence that its probable source was a current flowing from the northeast or north-northeast across the area. The faunas that formed the abundantly fossiliferous assemblage zones of the lower Trail City intermittently occupied a lobate area of about 800 to 1000 square miles trending slightly southwestward across the area and terminating in that direction within it. As the distribution maps show (Fig. 19) the position and outline of this lobate area of accumulation was approximately the same for all assemblage zones, indicating that conditions in this limited area periodically favored the establishment of a populous benthic fauna whose preservable elements were dominantly molluscs. The characteristically mixed sediment which occupies much of the Trail City sequence in this lobate area and extends at least a short distance marginal to it, forming the Little Eagle lithofacies, indicates the presence of a somewhat more persistent fauna of soft-bodied burrowers. Peripheral to the area, the relatively undisturbed, thin-bedded sediments of the Irish Creek lithofacies indicate much less favorable living conditions which supported a small infauna of protobranch bivalves and relatively few soft-bodied burrowers. The general coincidence of the Little Eagle assemblage areas with the path of growth of the succeeding Timber Lake sand body (Figs. 16 thru 23) unquestionably relates the productivity of the areas to the southwestward-flowing current. Independent evidence for the current within the Trail City is the small sand body that formed along the axis of the Little Eagle assemblage areas during the appearance of the *Protocardia-Oxytoma* assemblages (Fig. 19).

The coterminous molluscan assemblages of the lower Trail City imply rather static limiting conditions through the accumulation of the *Protocardia-Oxytoma* assemblages. The recurrent rather than continuous occupation of the area, however, indicates periodic variations in the environment within these persistent limits. From the fact that the Trail City Member grades into a sandy "Fox Hills" complex about 50 miles southwest of the Little Eagle lithofacies it follows that the landward side of the type area lay to the west. The little subsurface evidence due west of the type area is insufficient to indicate how far from the shoreline the accumulations lay. The presence of conspicuous subdominant associations along the west side of both the Lower *nicolleti* and *Protocardia-Oxytoma* Assemblage Zones probably reflects subtle changes shoreward, as does the conspicuous banding and sparse fauna of the Irish Creek lithofacies west of the Little Eagle lithofacies. The latter is so marked as to suggest that any topographic differentiation, achieved with the build-up of the Timber Lake sand body, might have been incipient in Little Eagle deposition, but there is no evidence for this.

The absence of abundant molluscan faunas in the Little Eagle lithofacies during the deposition of the upper Trail City is the only obvious difference between it and the lower Trail City. Except for the northern part of the type area, which was inhabited by the *D. abyssinus* fauna just in advance of the influx of Timber Lake sand, scattered protobranchs and a few ammonoids were the only obvious shelled inhabitants. The few thin beds of silt and sand, commonly with glauconite, that form the upper jarositic zone approximately mark the arrival of the sand body in the northeastern part of the area and introduce a second episode of occupation by shallow marine faunas, this time associated with the dominantly clayey sand fringing the marginal areas of the sand body.

The small part of the type area, shown on Fig. 20, in which nearly clay-free sand is in relatively sharp contact with parts of the Trail City approximately at or below the upper jarosite indicates that the axial part of the Timber Lake sand body grew longitudinally, as a relatively narrow, bar-like body, into the area from the northeast. As this body continued to grow, its direction changed slightly, becoming south-southwest. By the time the faunas of the *Sphenodiscus* concretion layer appeared (Fig. 21), the axial portion of the body had reached the center of the type area, all of its major component lithofacies were represented within the area, and sand had spread nearly to its maximum extent. Recognizable on the facies map at this level is the southwestward-tapering, axial part of the sand body with an area of coarser, cleaner, current-bedded sand on its presumably higher part. These latter sands, carrying the *Tancredia-Ophiomorpha* biofacies, mark the shallow, probably intertidal part of the sand body. The extent of the shallow facies at this level is obscured by poor outcrops, but is inferred to have crossed the Grand Valley because of the extent of the thin-bedded Rock Creek lithofacies that lies immediately to the west of the axial part of the sand body. The thin-bedded sands and sandy clays of the Rock Creek lithofacies are most likely a subtidal barrier flat facies which grades westward into the thin-bedded clays and silts of the Irish Creek lithofacies, here representing the bay environment behind the growing barrier.

To the south and southeast of the slightly emergent axial part of the sand body, clayey sands spread widely. The south-southwest direction of growth of the sand body across the west end of what is now Dewey County, is indicated by the presence of local sand lenses in the sandy clays in this area. The fringe of clayey sand extends eastward to the end of outcrop, becoming more clayey in this direction. West of the projected axis of the sand body it grades abruptly into the Irish Creek lithofacies.

Faunas at the *Sphenodiscus* level (Fig. 21) show greatest diversity in the clayey sands down-current south-southwest of the well-defined axial part of the sand body. Here, in the Moreau Valley in westcentral Dewey and adjacent parts of Ziebach Counties, the *D. abyssinus* association occurs in scattered concretions in and about the concretion layer of the *Sphenodiscus* level. *Sphenodiscus* and a small associated fauna of bivalves occur in the latter concretions. *Cucullaea* is locally present in this fauna south of the Moreau River. The less diverse and generally sparser fauna dominated by *Pteria linguaeformis* and *Ostrea pellucida* occupies the clayey sands east of the axial part of the sand body and the area of greater faunal diversity that lies in its longitudinal projection.

Growth of the sand body from the level of the *Sphenodiscus* concretion layer through that of the *Cucullaea* Assemblage Zone is largely subtidal, although the axial area of very shallow sands with *Tancredia* and *Ophiomorpha* reach the middle of the Moreau-Grand divide and begin to spread laterally (Fig. 22). The subtidal sands spread throughout the area previously occupied by clayey sand, and areas of the latter are found only on the eastern tips of the divides. The west side of the sand body holds its position, except for a very slight westward shift in the Grand Valley area. Deposits west of the sand body are still the silty and sandy gray clay of the Irish Creek lithofacies.

The marked northeastward spread of the more diverse marine faunas now characterized by an abundance of the thick-shelled bivalve *Cucullaea* may be a result of the stability of the sand body during the interval in question. At the level of the *Cucullaea* Assemblage Zone (Fig. 22) the diversity gradient from the deeper to the shallower parts of the sand body is shown more clearly than at any other level. The three major faunal associations present, in order of decreasing diversity, are the *Cucullaea* association, the *Cymbophora-Tellina* association and the *Tancredia-Ophiomorpha* association. Of the local variations, or gradations, the more conspicuous is the *Dosiniopsis-Tancredia* association that lies on the west side of the axial *Tancredia-Ophiomorpha* biofacies, and is gradational between it and the *Cymbophora-Tellina* association. *Tancredia* also occurs in the *Cymbophora-Tellina* association west of the axis of the sand body but not east of it.

The principal exception to the general paucity of fossils in the Irish Creek lithofacies is the presence of fossiliferous concretions of the *Cucullaea* Assemblage Zone. In this lithofacies the northward change from a *Cucullaea* association to a *Cymbophora-Tellina* association matches the change in the adjacent sand body. Within the type area the continuity of these faunas across a well-defined lithofacies boundary has little obvious effect on the species except that specimens of *Cucullaea* are predominantly smaller in the clayey Irish Creek.

The maximum spread of the *Cucullaea* association, reached at the level of its assemblage zone, marks the initiation of renewed growth in the sand body. By the level of the *Cymbophora-Tellina* Assemblage Zone (Fig. 23), a well-defined axial part of the sand body is no longer recognizable as the shallow sands with the *Tancredia-*

Ophiomorpha association have spread widely in the northern part of the area, occupying all but a narrow strip along the west side of the sand body. Although this broad area of shallows tapers somewhat southward it apparently terminated along a fairly broad southern front, now obscured in the poor exposures along the south side of the Moreau-Grand divide east of the Little Moreau River. Additional spread of sand westward in the Grand Valley area straightens the abrupt west side of the sand body to nearly a due north-south trend.

From its maximum extent the *Cucullaea* association shifts southward and is replaced by the *Cymbophora-Tellina* association, with only a small area on the Cheyenne-Moreau divide apparently preserving a *Cucullaea* association at the level of *Cymbophora-Tellina* Assemblage Zone.

Cessation of the longitudinal advance southward of a well-defined axial part of the sand body implies a change in regimen. This coincides with the appearance of the Bullhead lithofacies on the west in the area formerly occupied by the Irish Creek lithofacies. At the level of the *Cymbophora-Tellina* Assemblage Zone only a narrow area, a few miles wide, of Irish Creek lithofacies lies between the west edge of the Timber Lake sand body and the eastward-advancing Bullhead lithofacies. The appearance of the Bullhead just above the *Cucullaea* Assemblage Zone over most of the western part of the type area, and beyond it to the southwest, implies rapid spread of this thin-bedded, shallow-water marine facies. Although the Bullhead and Colgate do not comprise a lagoonal facies in the strict sense, they are in effect the filling of the shoreward area west of the sand body during the latter part of its development.

From the patchy record that remains of the Timber Lake sand body above the *Cymbophora-Tellina* Assemblage Zone, it is evident that the shallow *Tancredia-Ophiomorpha* sands occupied all of the body in the northern part of the area, and spread only slightly farther southward. Along the northern breaks of the Moreau, *Tancredia* and *Ophiomorpha* appear locally associated with *Ostrea, Pteria, Cymbophora* and *Panope* in some of the highest preserved exposures of the Timber Lake, over 50 feet above the level of *Cymbophora-Tellina* zone. This fauna is identical to that bordering the *Tancredia-Ophiomorpha* biofacies at the *Cymbophora-Tellina* zone level on the west side of the sand body in the Grand River area; it suggests that the intertidal area of the sand body may never have extended south of the latitude of the Moreau River. This is in keeping with the fact that it has not been found on the Cheyenne-Moreau divide, where only more diverse marine faunas occur, although here there are no exposures in which the undoubted top of the Timber Lake Member can be seen.

Apparently the change in the regimen of deposition of Fox Hills sediments indicated by the abrupt appearance of the Bullhead lithofacies on the west, slowed and soon terminated the axial southward growth of the shallow, probably intermittently exposed, top of the Timber Lake sand body. It did continue to enlarge laterally, however, and sand accumulated in subtidal areas off its south end and east side. In its late stage in the type area, the sand body was most likely the south end of a low, largely submerged bank. The appearance of the Bullhead lithofacies in the area west of the sand body at approximately the same time that it ceased to grow southward suggests increased energy of currents entering the area from the west, concomitant with decreased energy of the coastal currents responsible for the growth of the sand body. Eventually, this differential increased to the point where the thin-bedded Bullhead deposits spread laterally at the expense of the Timber Lake sand body and overlapped a large part of it. The Bullhead lithofacies spread over most of that part of the sand body capped by the *Tancredia-Ophiomorpha* biofacies in the northern part of the type area and at least part of the subtidal sands with the *Cymbophora-Tellina* association on the west side of the sand body in the Moreau Valley. If it covered the entire type area the record is lost due to erosion throughout the Cheyenne-Moreau divide and parts of the south side of the Moreau-Grand divide. In the latter area, certain indurated butte caps which rest directly on the Timber Lake sand may indeed be Colgate lithofacies; if so, they most likely indicate the presence of raised areas on the thicker part of the Timber Lake sand body that were not covered by sediments of the overstepping Bullhead lithofacies.

Bullhead and Colgate lithofacies represent very closely related environments, the latter occurring interbedded or as channels in the presumably shallow-water Bullhead. The brownish-gray color of the latter suggests the possibility of intertidal exposure but there is no evidence that these thinly-interbedded silts, sands and silty clays were tidal-flat deposits. From the contained faunas in the southwestern part of the type area it seems more likely that their environment was subtidal.

Lenticular bodies of Colgate lithofacies, and even some of the Bullhead lithofacies high in the section, carry partings or thin layers of lignitic material. Relationships at the upper contact of the Iron Lightning Member suggest gradation, with only local unconformity, into the salt marsh and coastal plain deposits of the Hell Creek Formation (Pl. 11, figs. B and C). Lenses of Colgate-like sand are locally common in the lower Hell Creek, but equally common are channels cut into either Colgate or Bullhead lithofacies and filled with clays and lignitic clay. Locally, estuarine silty clays with concretionary masses of the *Corbicula-Crassostrea* association also occur in the lowermost Hell Creek. Presumably the Fox Hills throughout the type area was succeeded gradationally by these coastal plain deposits. Although lateral gradation of most or all of the Iron Lightning Member into Hell Creek is indicated to the west and southwest, in the type area little or none of this is apparent.

ENVIRONMENTAL UNITS

Within the well-established pattern of interior Cretaceous sedimentation the Fox Hills Formation has long been recognized as a variable unit of shallow, marginal marine deposits formed during the filling up and/or shallowing of the basin of marine deposition. The three members of the type Fox Hills and their various lithofacies permit a fairly clear reconstruction of the nature of the marginal environments in the type area. Figure 12 summarizes diagramatically the principal Fox Hills lithofacies within the area and indicates their spacial and stratigraphic relationships.

The Trail City Member marks an abrupt change in sediment type from the relatively homogeneous finely silty clay of the underlying Elk Butte Member of the Pierre. Apart from its one large body of organically mixed sediment (Little Eagle lithofacies) it is a thin-bedded to laminated clayey silt and silty clay. The silt and fine-sand content is greatest in the Little Eagle lithofacies and, again, in the western part of the Irish Creek exposures along the Moreau River. Source of sediment could have been from both the northeast and the west, with the former the dominant.

Lamination and thin-bedding are not by themselves diagnostic. As Van Straaten (1959, p. 214) has illustrated, primary depositional factors tend to produce lamination and cross-lamination in most shallow and neritic environments; whether the lamination remains or is destroyed by burrowing organisms "depends on the ecological conditions and on the rates of deposition or of reworking by waves (and/or currents)." The lower part of the Trail City containing the populous settlements of organisms thickens southwestward into the laminated Irish Creek lithofacies; whether ecological conditions or rate of deposition controlled the distribution of facies during this interval is obscure. The fossil assemblages and the mottled beds between them indicate that sedimentation was not too much for soft-bodied burrowers and at least periodically permitted growth of large molluscan populations. Moreover, the coterminous Assemblage Zones suggest stability in the current system and until shortly after the last of these (*Protocardia-Oxytoma*) there is little evidence of growth in the approaching sand body. The somewhat thicker Irish Creek beds may indicate a proportionately greater influx of sediment from the west; if they are dominantly derived from the northeast there must have been by-passing in the area of the assemblages. There is some evidence of the latter in the northwestern part of the lower Little Eagle lithofacies where one finds general thinning of the interval, omission of the Upper *nicolleti* Assemblage Zone, and convergence of concretion layers in the *Limopsis-Gervillia* Assemblage Zone. For the lower Trail City, then, there is evidence that the northeastern sediment source was dominant, but whether the distribution of benthic faunas was controlled by rate of sedimentation or some other environmental factor is not evident.

During deposition of the upper Trail City the growing sand body entered the area and the interval is obviously one of domination by the northeastern current system. Lamination in the Irish Creek beds to the west is more spotty in distribution and the interval here is about the same thickness as in the Little Eagle lithofacies. Growth of the sand body appears continuous until its emergent end reaches the approximate center of the type area; gradual coarsening of sediment during this growth resulted in the gradational nature of the Trail City and Timber Lake members over most of the area. The submarine apron of clayey, silty sand of this transition changes very abruptly on its west side to the darker sandy clay of the equivalent Irish Creek. Areas of lamination occur only locally in the submarine apron of the sand body but are dominant in the Irish Creek. Laminations and thin beds throughout the Irish Creek vary somewhat in their regularity; in general, the thin beds have somewhat irregular surfaces though rarely can they be classed as nodular.

Another period of apparent stability of the sand body coincides with the fossiliferous succession of beds starting with the *Sphenodiscus* concretion layer and continuing through the *Cucullaea* Assemblage Zone. In both the sand body and equivalent Irish Creek lithofacies lamination is only locally present in this interval.

Subsequent to the deposition of the *Cucullaea* Assemblage Zone a marked change in depositional pattern takes place, and the interval between the latter zone and the horizon of the *Cymbophora-Tellina* Assemblage Zone is characterized by 1) lateral spread with very little southward growth of the sand body, including broad spread of its uppermost intertidal facies, and 2) the appearance of the Iron Lightning Member throughout the western part of the area. The latter, a spectacularly laminated sequence of clay silt and sand in various mixtures, is everywhere in sharp contact with the Irish Creek lithofacies. Because the uppermost Irish Creek is not well laminated in exposures where the contact can be seen, it has not been possible to determine whether or not their respective laminae are parallel. Laminations of the Iron Lightning Member are dominantly parallel laminations and contrast rather conspicuously in

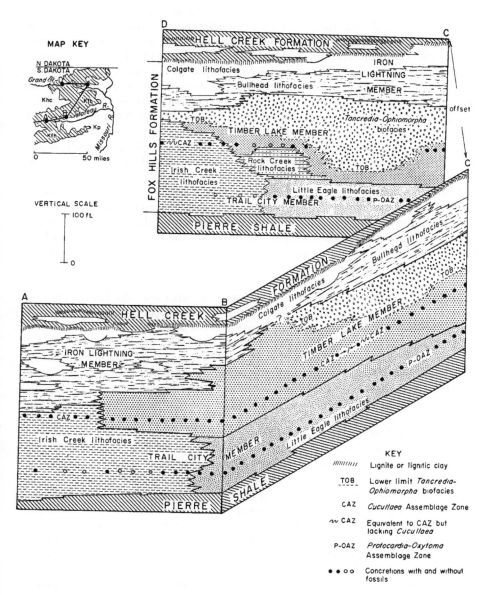

FIG. 12. Diagrammatic summary of the relationships of the Fox Hills lithofacies in the type area.

structure and in their lighter gray to brownish gray colors with the laminated parts of the Irish Creek. Mottling, locally present in the Irish Creek is conspicuously absent in the Iron Lightning Member.

The abrupt appearance of the Iron Lightning Member in the sequence coincident with a change in the growth pattern of the sand body suggests the introduction of a different depositional regime from the west whose currents became dominant as the northeastern current waned. On the other hand, outcrops in parts of the area show a succession in which the Iron Lightning Member is in sharp contact with the underlying shallow beds of the Timber Lake sand body at different levels; this widespread relationship on the sand body is suggestive of a progradational barrier-lagoon sequence, implying the progressive development of a single rather common type of marginal deposition. These alternatives mark a critical point in the environmental interpretation of the Fox Hills; with careful selection of exposures, a strong case could be made for the barrier-lagoon interpretation, but it does not hold up when the successional pattern over the entire type area is examined.

If the lithofacies of the Iron Lightning Member are lagoonal deposits the seaward barrier should be a persistent flanking facies on the east. But the stratigraphic evidence indicates that the Timber Lake sand body built up to sea level only as far southwest as the central part of the type area yet the lithofacies of the Iron Lightning Member extend throughout the western part of the type area and many miles beyond it to the southwest (Fig. 13). The Iron Lightning Member did not form landward of the submarine part of the sand body anywhere in the type area so there is little likelihood that it did to the southwest. In the northern part of the type area the barrier-flat Rock Creek lithofacies of the Timber Lake sand body is associated with a well-defined barrier with an emergent top; yet it grades westward into the Irish Creek, not the Iron Lightning lithofacies. Nowhere in the area does the Iron Lightning appear in the sequence *before* the level of the *Cucullaea* Assemblage Zone in spite of the fact that the Timber Lake sand body was essentially completed before this.

The Iron Lightning Member also appears widely where there is no underlying Timber Lake sand body. Of course the possibility exists that the Timber Lake is the initial body in a contiguous, eastward-migrating progradational series, but erosion in the Missouri Valley has destroyed all evidence of what happens to the Fox Hills east or southeast of the type area. Moreover, it can be demonstrated on both the Moreau and Grand Rivers that Iron Lightning lithofacies between the horizons of the *Cucullaea* and *Cymbophora-Tellina* Assemblage Zones pass eastward *not* into the Timber Lake sand body, but into a long narrow body of Irish Creek lithofacies, several miles wide, that flanks the sand body on the west. This relationship is shown diagrammatically in Fig. 12. During this interval a profile across the area at the latitude of the Moreau River would have shown the sand body as a submarine barrier complex, the Irish Creek as deposits in a shallow trench behind it, and the Iron Lightning as shallow platform deposits grading to coastal swamp on the west.

Above the level of the *Cymbophora-Tellina* Assemblage Zone, the Iron Lightning, in the Bullhead lithofacies, is in direct contact with the Timber Lake sand body. Although the contact between the two is sharp, the basal beds of the Iron Lightning commonly include sand obviously derived from the Timber Lake, and at some places sand beds up to 2 feet in thickness, which is exceptional in the Bullhead lithofacies, occur just above the contact. At several localities Colgate lithofacies appear to be in contact locally with the Timber Lake sand. Relationships between these two lithofa-

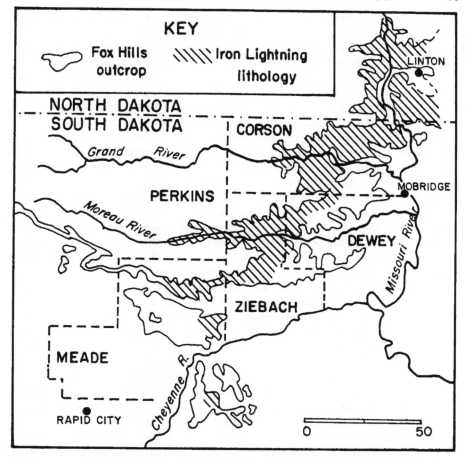

FIG. 13. Approximate distribution of outcrops of Iron Lightning lithology in and adjacent to the type area of the Fox Hills Formation.

cies are confused by localized induration of sand, but there is strong indication that the lower levels of Colgate lithofacies in the Iron Lightning Member come in contact with the Timber Lake sand where the former thin abruptly onto the sand body. Relationships between the Colgate and Timber Lake sands need elucidation; their contact may be more in the nature of an emplacement of Colgate in depressions scooped out of the Timber Lake than of a gradation of one into the other. Selective induration shown in the Timber Lake at some places (Pl. 5, fig. C) may actually have been confined to local lenses of Colgate in the top of the Timber Lake sand. In summary, there is no stratigraphic evidence of gradation between the Iron Lightning sediments and the Timber Lake sand body; interbedding near the contact could have resulted from local reworking, and the contact between the two members is everywhere sharp and parallel with the bedding in the Iron Lightning.

The Iron Lightning Member as a whole appears to have come from a westerly direction, filling in most of the area behind the Timber Lake sand body during a late stage of its deposition and eventually overstepping it. The known drop in section of the Fox Hills southwest of the type area as it passes into the atypical sandy, coal-bearing facies of the Stoneville area suggests that the western depositional regime represented by the Iron Lightning Member in the type area is an eastward-growing delta platform. The sedimentary structures of the Iron Lightning Member support this, as they are comparable to those in present-day delta-front environments.

Characteristics of the Bullhead lithofacies which indicate rapid deposition are laterally persistent, primary, regular layering and the paucity of organisms— indicated both by the absence of mottled zones and very sparse distribution of shells. The stratigraphic relations just reviewed attest to the overwhelming of the Timber Lake sand body by the Iron Lightning sediments, and this, together with the extent and thickness of the Iron Lightning Member, support the relatively rapid influx of a considerable volume of its sediment. Bullhead layering, or lamination, is directly comparable to that described from both topset and some foreset parts of present-day deltas. In the varying terminology of these environments, cores taken from areas in the Bullhead lithofacies such as those shown on Pl. 9, fig. B, and Pl. 10, fig. C, would compare with those shown and/or described from the Mississippi delta front platform by Shepard (1960, p. 67, fig. 7, C and D). Moore and Scruton (1957, p. 2727, fig. 3), and others. As Van Straaten (1959, p. 208) pointed out, distinct laminations also occur to varying depths on the foreset slope of different deltas but are apparently common to most deltas only on the upper parts of the slope—the "proximal fluviormarine" deposits of Rhone delta workers or upper "pro-delta silty clays" of Mississippi delta workers. Allen (1964, p. 31) noted that in the Niger delta deposits of both the delta platform and shallower pro-delta slope are well laminated. Van Straaten (1959, p. 208) lists angle of foreset slope, current velocities of rivers and average composition of sediment as factors that account for the varying extent of laminations down foreset slopes of deltas. As any delta in the vicinity of the type Fox Hills would necessarily have very slight gradient on the foreset slope and have the sediment composition shown on the cumulative curves of samples from the Bullhead lithofacies (Fig. 10), it is reasonable to expect marked lamination of the foreset slopes. On the basis of lamination, then, the Iron Lightning sediments could either be delta front platform, or pro-delta slope deposits, or a combination of the two. The great extent and continuity of the laminated Bullhead lithofacies tends to eliminate other, more specialized environments in which regular lamination might be expected to occur.

The Colgate lithofacies, which appears within the Iron Lightning Member at several levels west of the area occupied by the Timber Lake sand body but only at the top of the member over the sand body, occurs as channel-fills, beds, and irregularly shaped lenses which grade abruptly laterally into the thin-bedded or laminated clay, silt and sand of the Bullhead. Fragmental plant matter abounds in both the Bullhead and Colgate but is more common and coarser in the latter. Slump structure is common in the Bullhead adjacent to bodies of Colgate sand and locally involves the latter. Bedding in the Colgate sand ranges from regular lamination to trough cross-lamination, and composition from sandy clay to medium-grained sand. In their special features, relationships with the Bullhead lithofacies and, at the top of the Iron Lightning Member, with the coastal plain deposits of the overlying Hell Creek Formation, the Colgate sands are similar to the coarser, topset delta-platform deposits

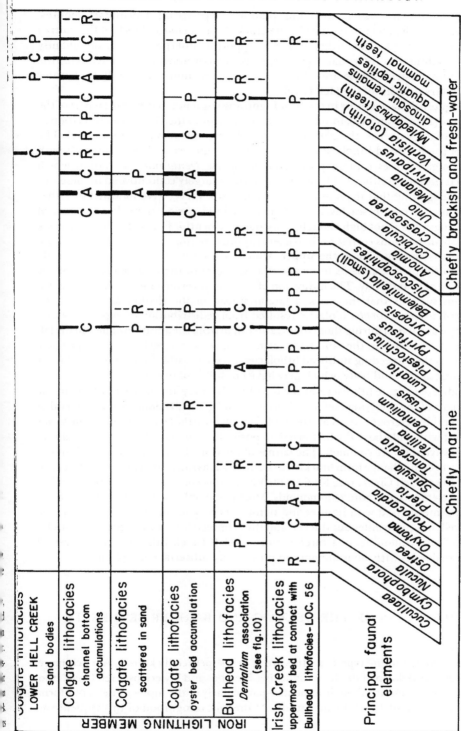

FIG. 14. Marine-continental transition illustrated by the distribution of the principal fossils found in the upper Fox Hills and lower Hell Creek of the type area. Letters indicate apparent abundance: R = rarely present, P = present, C = common, A = abundant.

of the distributary system, including platform deposits off distributary mouths. Faunas of the channel fills typified by the brackish *Corbicula* and reefs of *Crassostrea* are unique to the Colgate lithofacies and support its interpretation as dominantly shallow subtidal to intertidal topset deposits in environments associated with the distributary system. These are, of course, continuous into dominantly fresh-water distributary deposits of the Hell Creek.

Colgate lithofacies recur in the Iron Lightning sequence in the western part of the type area (Figs. 10 and 12), indicating that twice prior to the Hell Creek transition at the top of the sequence, the growing pile of sediment was very close to sea level. The beds of the Bullhead lithofacies immediately overlying each of these intervals with local Colgate lenses contain pockets of the sparse *Dentalium* association which characterizes the Bullhead. The Bullhead fauna is largely concentrated in these two layers, suggesting that these initial approaches to sea level were followed by either subsidence or rise in sea level and that for a brief period thereafter the rate of sedimentation was not high enough to discourage a benthic fauna. The lower of these two Colgate horizons corresponds to a level approximating that of the *Cymbophora-Tellina* Assemblage Zone; the upper, which is thicker, more varied and locally contains *Corbicula,* appears to be about at the level of the uppermost beds of the Timber Lake sand body. The indication of eustatic lowering of sea level at those two horizons suggests that the marked lateral spread of the shallow sands with the *Tancredia-Ophiomorpha* biofacies resulted from redistribution of exposed barrier sand by wave action during those times rather than from the addition of new material. At the second horizon the sand body seems to have been largely leveled and transgressed by the succeeding Bullhead lithofacies, except possibly for parts of the thicker axial area of the sand body, which may have no Bullhead over it.

Bodies of Colgate lithofacies at the top of the Iron Lightning Member and in the lower part of the overlying Hell Creek occur in and generally dominate an interval of considerable heterogeneity. Variations of the Colgate lithofacies dominate the lower part but with and above the first marsh deposits, lignitic clays, lignites, and bentonitic, massive clays are common. The faunas of the Colgate channel sands at different levels reflect the change from brackish to fresh water, as illustrated in Fig. 14. But even in the lower Hell Creek, rare local estaurine deposits contain *Crassostrea* with a few marine elements, including ammonoids. More commonly, bodies of Colgate lithofacies are interspersed with lignites and lignitic clays. Some of the lignite beds cover hundreds of square miles but thicken and thin appreciably and are locally cut out by channelling. Undoubtedly a careful study of Hell Creek lithofacies would reveal many analogues with environments on present-day subaerial delta plains.

RELATIONS OF THE DELTAIC AND BARRIER SEQUENCES

Deltaic deposits in the type Fox Hills are at best superficially known. A much more complete breakdown of the Iron Lightning Member and lowermost Hell Creek into the numerous deltaic subenvironments—particularly of the topset or delta platform part—is anticipated with further study. At present, recognition of this sequence as deltaic is based on 1) the stratigraphic relations with adjacent units to the west of the type area, 2) the succession of structural types in the Iron Lightning Member and its close analogy with present-day deltaic successions of proximal fluviomarine deposits, 3) the presence of abundant plant remains, mica, and ferruginous aggregates, and 4) the sparse faunas of low diversity whose associations are consonant with the interpretation of environments based on sedimentary structures and stratigraphy.

Acceptance of the deltaic nature of the Upper Fox Hills-Lower Hell Creek sequence in the type area raises the question of its relationship to the barrier sequence of the Lower Fox Hills. In relatively shallow epicontinental seas whose coasts, in late phases of regressive cycles like the one in question, did not feature abrupt dropoffs into offshore basins, the burial of coastal marine deposits by sediments of broad shallow deltas consisting chiefly of proximal fluviomarine environments may have been fairly common. Botvinkina and Yablokov (1964, p. 41) note that the delta deposits of the Carboniferous coal measures in the Donetz Basin are "characterized by their superposition, often even with gradual transitions, on typically marine deposits." In the type Fox Hills the sandy barrier facies and the deltaic deposits that overstep it contain sands of almost identical mineral composition and size range, the chief differences being in the greater degree of decomposition of grains in the barrier sands and in their glauconite content. The barrier may well consist of sands reworked from abandoned areas of an adjacent delta to the north or even of an older abandoned area of the same deltaic complex. It may also be material carried from an actively growing delta to the north. The extent of the Iron Lightning lithology in the region (Fig. 13) indicates either a single very large delta or yoked deltaic deposits from more than one river system. The Iron Lightning crops out in a continuous belt approximately 150 miles along its environmental strike; by way of comparison, the shallow submarine deposits of the Orinoco delta are shown by Nota (1958, fig. 39) to extend along the coast more than 250 miles from the Orinoco mouths southeastward to the mouth of the Essequibo River.

For any of these possible sources of sand it seems necessary to invoke coastwise current to explain the growth of the barrier, although the abundance of plant remains —including such coarse material as the 8-foot *Palmoxylon* stems—indicates that a considerable source of fresh water contributed to it. Study of the sequence in the Missouri Valley area of North Dakota should answer some of these problems.

Relationships of the deltaic sequence to the Lower Fox Hills in the western part of the type area are more direct but little more edifying. The laminated Irish Creek lithofacies, parts of which thicken westward, could conceivably be derived from that direction and actually be a lower submarine portion of the advancing delta. The sharp contact between the Irish Creek and overlying Iron Lightning could be interpreted as the "very minor discordance" between topset and foreset beds, or between foresets and bottomsets, mentioned by Shepard (1964, p. 14) as one of the potential criteria for identification of the marine portions of deltas. Again the evidence for the extent of the deltaic sequence downward in the section lies outside the type area and the problem awaits further work. At present it is not known whether a foreshortened deltaic sequence of topsets and shallow foresets is superimposed on a submarine topography resulting from a completely separate depositional regimen or whether it is a more fully formed delta overriding a barrier that formed on its own bottomset or lower foreset slope during a period of depositional stability.

ASPECTS OF FOSSIL DISTRIBUTION

The many relationships between different fossil associations, and between fossils and lithofacies, revealed in the preceding description of Fox Hills stratigraphy, indicate that the patterns of fossil distribution reflect to a considerable degree the original distributions of organisms. If this is so, it should be possible to combine what is known about the ecology of different genera and species with the stratigraphic and sedimentary evidence and reconstruct the local Cretaceous environments in some detail. But the paleontological work on interior molluscan faunas, with the exception of the cephalopods, is antiquated and a critical revision of the systematics together with careful appraisal of the ecology of individual species based on functional morphology and associational data are prerequisite to any really definitive interpretation of the environment. Only the environmental implications of some of the more conspicuous aspects of the fossil assemblages are treated here or noted earlier in the text.

Perhaps the most obvious distributional feature of the Fox Hills faunas is that some occur in masses in limestone (calcitic) concretions and others are found loose in the matrix. Although examples of both modes of occurrence can be found in all members of the Fox Hills there is a great dominance of concretionary accumulations in the Trail City and lower Timber Lake members and a great dominance of "free" fossil accumulations in the upper Timber Lake and Iron Lightning members. Stated environmentally, the concretion occurrences are largely in the off-shore marine phase, specifically in and around the lower submarine parts of the barrier sand body. Most of the concretions in these facies resulted from the formation of calcite interstitially in the clayey silts and sands without apparent disruption of the sediment, indicating that they formed in sediment *subsequent* to burial of fossils. Crushed fossils outside the concretions and uncrushed ones within indicate solidification of concretions prior to compaction. Consequently the formation of the concretions *did not influence the original distribution of the organic remains,* it only served to insure their excellent preservation.

FAUNAS OF THE OFF-SHORE SETTLEMENTS

The fossil assemblages of the lower Trail City Member have been interpreted as resulting from recurrent mass mortality with relatively little disturbance of their natural distribution prior to burial. Subsequent work has revealed nothing to discourage this explanation, which is presented elsewhere (Waage, 1964) in detail and will not be repeated here. Instead, conditions under which the faunas lived, and died, are discussed in light of additional data.

The areas inhabited by the masses of molluscs that make up the assemblage zones of the lower Trail City, and help define its Little Eagle lithofacies, are informally referred to here as settlements. No suitable term is available in paleoecology for such areas, characterized by abundant fossil assemblages which accumulated approximately where they lived. Nor is a satisfactory term available from ecology, for "bank" has topographic implications and "patch" implies smallness. A settlement is a biogeographic area defined by relative abundance of organisms; it may embrace a number of different faunal associations, or communities, and many patches.

The contemporaniety of individual assemblage zones, and of individual layers of accumulation within it, is critical to their interpretation as settlements. None of the empirical data suggest the alternative—that the layers transgress time; several lines of evidence point to the contemporaniety of individual layers. It has already been pointed out that the assemblage zones and their component layers hold the same stratigraphic position relative to one another throughout their extent; other key beds such as the jarosite layers conform to this fixed pattern. Bentonites have been dispersed by organisms over most of the settlement area so that none can be continuously traced and shown to parallel the assemblage zones throughout their extent. But in marginal areas, concretion layers do hold a fixed stratigraphic position relative to bentonites, though with the possible exception of the D bentonite and underlying *Cucullaea* concretions, these areas are not broad enough to serve as conclusive proof of parallelism. One very strong indication of contemporaniety of layers is the fact that the base of the Timber Lake sand body, the one obviously time-transgressive feature in the area, cuts obliquely across the lower Trail City assemblage zones to the north-northeast along its longitudinal axis.

Some distributional features of the fossil assemblages themselves clearly imply contemporaniety of the layers. Preservation of the assemblages with relatively little change from their living arrangement is indicated by a number of characteristics (Waage, 1964, p. 556-559), two of which—the uniformity of faunal composition in individual layers and the numerical dominance of one or two species in each assemblage—are distinctive features of modern marine bottom communities (Thorson, 1957, p. 467). A fossil accumulation of marine benthic molluscs, like those in the lower Trail City, that is distributed continuously over a limited area and displays natural community structure is difficult to interpret as anything but a synchronous feature. In contrast, the *abyssinus* concretions found in the upper Trail City of the Grand River valley and in the lower Timber Lake along the Moreau Valley fortuitously provide a first-hand example of the mode of occurrence of a time-transgressive fossil association which is comparable in its general features to the lower Trail City associations and also occurs in the same general area. The *abyssinus* concretions do not form a continuous layer that cuts upward across the Trail City-Timber Lake contact from north to south; instead they form a number of much less extensive, disjunct layers distributed in different parts of the area at progressively higher horizons southward. The likelihood is extremely remote that a continuous, faunally uniform layer could form as the product of the gradual migration of a bottom community through time and space in the marginal areas of a sea, where conditions of energy and sedimentation are constantly changing.

A piece of negative evidence is afforded by 3 or 4 species of *Discoscaphites*, immigrants from the Gulf Coast region, which appear abruptly in the *Limopsis-Gervillia* Assemblage Zone and persist throughout the assemblage zones above, but except for one or two rare specimens of two of the species are not present in the Lower *nicolleti* Assemblage Zone below. If the assemblage zones were time-transgressive, immigrant species should be present in most assemblage zones in one part of the area and absent from most in another. Although no one piece of evidence indicating synchroniety of the assemblage zones is in itself conclusive, together they convincingly favor it.

Assemblage zones associated with the Timber Lake sand body have the same basic characteristics as those described for the Trail City zones (Waage, 1964, p. 553) but there are some differences. Characteristics common to all the assemblage zones include:

1. Great abundance of specimens with one or two species numerically dominant.
2. Excellent preservation of most specimens, the bivalves commonly with unseparated valves.
3. Random orientation and lack of size-sorting of specimens in individual concretions, but tendency for dominant bivalve species to occur in size-group aggregations.
4. Distribution over a limited area, the settlement, beyond which the horizon is unfossiliferous.
5. Dominance of one particular faunal association in a settlement with patterned distribution of subdominant associations relative to it.
6. Aggregation of individual species in clusters.

The first three characteristics are much the same in each settlement although there is somewhat more breakage of specimens in the Timber Lake, possibly owing to the somewhat higher turbulence of the environment. There is also a tendency for Timber Lake settlements to have more than two numerically dominant, though not always conspicuous, species.

The chief differences between Trail City and Timber Lake assemblage zones are in the last three characteristics, all of which are distributional features. The differences are of degree not of kind. Trail City settlements occupy a smaller area than those of the Timber Lake. In the lower Trail City the successive settlements show rather marked differences in dominant associations and in most, the conspicuous diversity gradient decreases from east to west within the type area. The *Limopsis-Gervillia* Assemblage Zone appears to have a north-south gradient but actually this is due to partial overlap of separate *Limopsis* and *Gervillia* settlements. Speden reports (oral communication) that the *Protocardia-Oxytoma* settlement has a dominantly protobranch association just northeast of the type area.

This forecasts the orientation of diversity gradients in the Timber Lake settlements, for in both the *Cucullaea* and *Cymbophora-Tellina* Assemblage Zones the principal diversity gradient decreases north-northeastward from the deeper downcurrent part of the barrier to its shallower, probably intertidal, central part (Figs. 22 and 23). The conspicuous difference in the faunas of these two zones results from the southward shift of the adjacent areas inhabited by the diverse *Cucullaea* and less diverse *Cymbophora-Tellina* Assemblages.

The rather faint east-west diversity gradients of the lower Trail City settlements (the *Protocardia-Oxytoma* settlement has east-west gradient to local *Lucina* patches and northeast gradient to protobranchs) contrast with the strong axial northeast gradients on the Timber Lake sand body. But when the Timber Lake assemblages are examined in more detail such distributional features as the occurrence of *Tancredia* and *Panope* on the west but not the east side of the sand body in the *Cymbophora-Tellina* Assemblage Zone show that there are east-west gradients as well, though they are poorly known and less conspicuous. Obviously, the two sets of gradients line up with the two depositional regimes; but it is not obvious that there is a causal relationship here. More likely the east-west gradient results from a difference in the seaward and landward sides of the northeast current and the barrier it deposited. In this connection it is interesting, if not significant, that the single echinoid specimen found in the type Fox Hills (the clypeasterid *Hardouinia*) came from the *Cymbophora-Tellina* zone on the easternmost tip of the Moreau-Grand divide, the seaward side of the sand body.

Clustering of species, which is so evident in the assemblage zones in the Trail City, is obvious chiefly in the *Cucullaea* Assemblage Zone in the Timber Lake Member. Here both *Cucullaea* and *Protocardia* commonly occur in clusters. Other Timber Lake fossil assemblages in concretions also show clustering but not as conspicuously. One of the most convincing examples that clustering is a natural feature of the settlements is reported by Speden (1965, unpublished) to occur in the *Cymbophora-Tellina* Assemblage Zone at a few localities on the north side of the Moreau River. Here *Cymbophora* and *Cucullaea*, usually mutually exclusive genera in the type area, occur together in the same concretion layer, but each only in clusters of its own species.

CONDITIONS FAVORING SETTLEMENTS

In attributing the origin of the assemblage zones to recurrent mass mortality (Waage, 1964, p. 562), it was pointed out that prevailing conditions during deposition of the Trail City may have been inhospitable to benthic organisms. The settlements would then represent relatively short periods of atypical, but hospitable conditions which were terminated by mass mortality on the return of prevailing conditions. This possibility must still be borne in mind for the Trail City settlements, but the Timber Lake settlements differ, there being at least a sparingly fossiliferous sequence from below the *Cucullaea* Assemblage Zone to and above the *Cymbophora-Tellina* level in the submarine part of the sand body. This is a distinct difference in the two intervals of assemblage zones (lower Trail City and lower Timber Lake) which suggests either some environmental difference or the advent of a few more tolerant species, like *Pteria* and *Cucullaea*, which seem to make up the bulk of the inter-assemblage zone faunas of the Timber Lake and are both very rare in the lower Trail City. But too little is known of the genera involved to evaluate this difference.

More significant is the fact that the two intervals with settlements correspond to periods of relative stability of the sand body, whereas the largely barren interval between is marked by the advance of the emergent part of the barrier into the type area. A second significant fact is that the Trail City settlements lie in the path of the sand body and are obviously controlled in their distribution by the northeast current. The lower Timber Lake assemblages, while more widespread, also have their richest faunas localized off the down-current end of the sand body. Taken together, these two sets of empirical data suggest that optimum conditions for growth of a settlement obtained only under the influence of the northeast current, but not while the emergent axial part of the barrier was being advanced. This implies that the current furnished one or more critical factors to the environment that were absent or in short supply in the surrounding environment, while the period of bar building introduced a detrimental factor or factors.

OTHER FAUNAS

Fox Hills fossil assemblages that do not occur in assemblage zones on submarine portions of the barrier include the *Tancredia-Ophiomorpha* association of the barrier top, the sparse *Dentalium*-dominated associations of the Iron Lightning Member, and the *Crassostrea* associations and channel deposits with *Corbicula* in the Colgate lithofacies. Some aspects of these faunas suggest particular environmental factors, but again too little is known of the animals themselves to place much reliance on their supposed ecology.

The sands with *Tancredia* and *Ophiomorpha* are more reliably tied to a specific restricted environment than are most Fox Hills lithofacies. Not only is the bedding structure of the sand unquestionably indicative of deposits in turbulent environments, but also their position on top of the barrier places them in the intertidal and/or supratidal region. *Ophiomorpha,* known for many years from empirical evidence to be restricted to marginal marine deposits and inferred to be a crustacean burrow (Brown, 1939; Häntzschel, 1952), has been rather convincingly attributed to the living marine decapod *Callianassa* or some closely related form (Weimer and Hoyt, 1964). In their study of *Callianassa* burrows at Sapelo Island, Georgia, Weimer and Hoyt found that the species is confined to littoral and shallow neritic sands from below mean low water to mean sea level (ibid, p. 763). Decapod remains have not, to my knowledge, formerly been reported from fossil *Ophiomorpha* burrows; like most mobile, infaunal invertebrates the animals probably surfaced at the onset of detrimental change in environmental conditions. However, one decapod fragment found in an *Ophiomorpha* tube in the Timber Lake Member (Pl. 8, fig. C) has been identified by Henry B. Roberts (U.S. Nat. Museum) as a callianassid, substantiating the Weimer and Hoyt inference that the Cretaceous structures were made by a member of this group. *Tancredia,* the principal bivalve associated with *Ophiomorpha* in the cleaner, shallow sands of the type Fox Hills, is an extinct, thick-shelled, infaunal genus with a tapered anterior and gaping posterior that indicate rapid burrowing and large siphons (Pl. 8, fig. B). It was obviously well suited to the shallow, turbulent environment, but is not as restricted as *Ophiomorpha,* occurring also with the *Cymbophora-Tellina* association in muddier sands west of the barrier as well as in a few other associations.

The faunas of the lower Iron Lightning Member were noted as being restricted largely to beds just above lower horizons of the Colgate lithofacies in the western part of the type area. These accumulations, probably in part current-formed, may represent periods of more than usual organic abundance for the lithofacies but this is difficult to verify. The fauna is odd in its local abundance of the small *Dentalium* and the otolith *Vorhisia,* sparse suite of infaunal bivalves, more common gastropods, and fairly numerous rostra of small belemnites. Concentration of the fauna in zones just above the Colgate lithofacies suggests that it lived during an early stage of inundation after the periods of shallowing indicated by the Colgate and before the rate of sedimentation in the delta front environment became intolerable. Modern *Dentalium* appears to require, or prefer, normal marine salinity; if this was also true of the Cretaceous form the thin layers must indeed have accumulated under prevailing marine waters rather than brackish water, although as Gunter (1947, p. 78) infers from information on fish, the small or the young of invertebrates are the more likely to venture into less saline water from normal marine salinities. This might apply to both the little belemnites and to *Dentalium*. That the water might not have been of normal marine salinity is suggested by the fact that the protruding tips of the little *Dentalium* shells are very commonly etched; but this feature could also result either from wear in sediment-laden current or from some kind of organic etching.

The more abundant oyster patch and channel faunas described from the Colgate lithofacies are distinctly different in faunal composition from other Fox Hills assemblages and obviously represent brackish environments. The *Crassostrea* association is restricted to the uppermost bodies of Colgate lithofacies in the type Fox Hills, and to local estaurine deposits in the lower Hell Creek. The presence of abundant, predatory marine gastropods has been suggested (Waage, 1967, p. 265) as a possible factor in this restriction of *Crassostrea* to brackish facies in the area of the type Fox Hills. Apparently optimum conditions for *Crassostrea* were similar to those for *Corbicula*, though the latter is more common and more widespread in the Colgate and probably could survive in areas of shifting sand. The presence of these two undoubted brackish-water indicators at the very top of the Fox Hills, and locally just within the first swamp deposits of the Hell Creek, must mean that normal marine salinity occurred very close to the emergent delta topset environments.

Channel deposits of Colgate lithofacies are of interest in that they show, in successive levels, the change from brackish to fresh water. Although a few broken and worn shells of *Unio* are present with an abundance of plant and nonmarine vertebrate remains in the Colgate channel deposit noted at the type locality of the Iron Lightning Member (Section 13, unit 12), *Corbicula* is the only dominant bivalve and it is relatively well preserved and occurs in great abundance. About 50 feet above the basal part of this channel deposit another is present locally, in the basal Hell Creek beds (Section 13, unit 20), in which whole bivalved shells of *Unio* are the only mollusc remains occurring with assorted bone fragments and teeth. Brackish-water fossils in the type area are thus restricted chiefly to the Colgate lithofacies of the Fox Hills. In terms of geographic distribution this must have been a rather narrow irregular zone including distributary mouths at the margin of the subaerial part of the delta. Considering the amount of fresh water that must have been flowing into the shallow sea the brackish zone seems narrowly restricted.

ENVIRONMENTAL FACTORS

Of the principal environmental factors that control the distribution of organisms, temperature can probably be eliminated in dealing with the relatively small area of the type Fox Hills. Dorf (1942, p. 100-103) evaluated the flora of the type Lance Formation as indicating a humid lowland with more nearly warm temperate than subtropical climate. A similar climate can be assumed for the type Fox Hills, which is correlative with at least the lower part of the type Lance, and does not lie very far from it. Under these conditions one might expect wet and dry seasons. The great amount of plant material in the Fox Hills suggests in itself at least periodic wet spells and river flooding. Charcoal fragments, which are locally abundant in the Colgate lithofacies, suggest burning-off of swamp vegetation in coastal areas, possibly in dry spells, but there is no concrete evidence of wet and dry periods.

Major environmental factors critical to the local distribution of Fox Hills faunas include salinity, oxygenation, turbidity and/or rate of deposition, and food. There is no way to evaluate the factor of food; nothing is known about the plankton or the algal growth. The considerable organic matter in the sediment may play an important role in the distribution of bivalves (Bader, 1954) and cannot be discounted as a possible factor in the localization of settlements, and in the general concentration of rich molluscan faunas on and around the Timber Lake sand body.

Variation in each of the factors of salinity, oxygenation and rate of deposition, taken singly, can be called on to influence the growth of, or destroy, the settlements and otherwise account for Fox Hills distributions, but there is no evidence that any one of these factors was dominant. However, the general low diversity of all interior Cretaceous marine faunas when compared with their Coastal Plain counterparts certainly suggests the pervading effect of a major environmental factor, such as salinity. In evaluating a local area like the type Fox Hills the spectre of this widespread unknown makes speculation especially discouraging.

The environmental facts of the Fox Hills biostratigraphy fit in with the assumption that the sea during Fox Hills time was normally saline. Certainly it was not of low salinity, for the distribution of *Crassostrea* and other known brackish elements is restricted very much as their counterparts are today and many elements of the marine faunas are not likely to have survived in waters of low salinity. General hypersalinity is a possibility, though one not supported by the presence of any of the saline deposits usually associated with local extremes of this condition. However, a slightly hypersaline sea and an abundant local influx of fresh water would not likely give rise to deposits of salts in coastal regions. Although this combination seems highly unlikely it would explain the rather restricted brackish zone and one might also account for the Fox Hills settlements in the paths of currents by postulating influx of enough fresh water from nearby deltas during floods to periodically bring about normal salinity in these shallow marine areas with resultant optimum growth. Shaw has argued (1964, p. 17-20) for the hypersalinity of epeiric seas on theoretical grounds. Athough the interior Cretaceous sea does not fit his model of an epeiric sea in many ways, his discussion of the development of high salinities in marginal areas of such seas, particularly during regressive phases, is pertinent to the problem of Fox Hills environments. On the empirical evidence at hand, which does not include geochemical data, the possibility of hypersalinity in Fox Hills marine environments cannot be refuted with much more conviction than it can be supported.

Assuming normal conditions of salinity the distribution of the faunas in the upper Fox Hills is fairly straightforward, with intertidal and shallow neritic elements on the barrier and a good brackish fauna in distributaries at the emergent delta edge. Scarcity of delta front faunas can be attributed to rate of sedimentation for it coincides with a known acceleration of delta growth.

The difficult distributions to explain are those of the settlements in the Lower Fox Hills. Little can be added to previous discussion of these, and the mass mortality they indicate (Waage, 1964, p. 560-562), until more is known about the habits of the organisms involved. Compared with the relatively barren upper Pierre Shale below, one can speculate that more food and better oxygenation was available in the current systems of the generally more turbulent regimes of Fox Hills sedimentation. Diversity gradients of the settlements are unquestionably related to some aspect of the regimen of barrier deposition. The fact that the two intervals containing settle-ments (lower Trail City and lower Timber Lake) correspond with slack periods of barrier growth suggests a delicate balance of optimal conditions between the benefi-cial factors associated with the currents and the inherent detrimental effects of excessive turbidity and rate of deposition of sediment.

SUMMARY

Sediments forming the Fox Hills Formation in its type area are products of two depositional regimes, a coastwise current from the northeast and drainage from a deltaic front advancing from the west. The former dominated sedimentation during accumulation of most of the lower part of the formation. At this time a barrier sand body advanced longitudinally into the area from the northeast, its submarine part crossing the area, its emergent part reaching only the center. Intermittent, populous settlements of dominantly molluscan benthos formed in the clayey silts off the down-current end of the sand body. Later, similar settlements formed on the sub-marine south end of the body, their faunas becoming less diverse northward around the emergent axial part of the barrier, which supported a very restricted fauna in the cleaner sands of its high subtidal to intertidal environment. West of the barrier, dark clayey silt and fine sand was deposited, largely in an irregularly thin-bedded sequence little disturbed by the scant fauna present until one of the molluscan settlements on the sand body spread over it. Shortly thereafter this back-barrier area began to fill from the west with clay, silt and sand in parallel thin beds and laminae, the eastward-growing shallow platform deposits of a large deltaic complex. From this point the deltaic sedimentation dominated, filling the back-barrier embayment and overstepping the barrier. The delta-front deposits and the sands of environments in and adjacent to distributaries, carrying brackish-water faunas, form the upper part of the Fox Hills and are in turn succeeded by the swamp and coastal plain deposits of the subaerial part of the delta which form the lower part of the Hell Creek Formation.

REFERENCES CITED

PUBLICATIONS

Allen, J. R. L., 1964, Sedimentation in the modern delta of the River Niger, West Africa, *in* Proc. 6th Internat. Sed. Cong.: Devel. in Sedimentology, v. 1, p. 26-34.

American Commission on Stratigraphic Nomenclature, 1961, Code of stratigraphic nomenclature: Am. Assoc. Petroleum Geologists Bull,. v. 45, p. 645-665.

Bader, R. G., 1954, The role of organic matter in determining the distribution of pelecypods in marine sediments: Jour. Marine Research, v. 13, no. 1, p. 32-47.

Bartram, John G., 1937, Upper Cretaceous of the Rocky Mountain area: Am. Assoc. Petroleum Geologists Bull., v. 21, no. 7, p. 899-913.

Birkelund, Tove, 1965, Ammonites from the Upper Cretaceous of West Greenland: Meddel. om Grønland, v. 179, no. 7, 192 p., 49 pls.

Böse, Emil, 1927, Cretaceous ammonites from Texas and northern Mexico: Univ. Texas Bull., no. 2748, p. 143-357, 18 pls.

Böse, Emil and Cavins, O. A., 1927, The Cretaceous and Tertiary of southern Texas and northern Mexico: Univ. Texas Bull., no. 2748, p. 7-142.

Botvinkina, L. N., and Yablokov, V. S., 1964, Specific features of deltaic deposits in coal-bearing and cupriferous formations, *in* Proc. 6th. Internat. Sed. Cong.: Devel. in Sedimentology, v. 1, p. 39-47.

Brown, R. W., 1939, Fossil plants from the Colgate member of the Fox Hills sandstone and adjacent strata: U.S. Geol. Survey Prof. Paper 189-I, p. 239-275.

Calvert, W. R., and others, 1914, Geology of the Standing Rock and Cheyenne River Indian Reservations, North and South Dakota: U.S. Geol. Survey Bull. 575, 49 p.

Cobban, W. A., 1958, Late Cretaceous fossil zones of the Powder River Basin, Wyoming and Montana, *in* Wyoming Geol. Assoc. Guidebook 13th Ann. Field Conf., Powder River Basin, Wyoming, 1958: p. 114-119.

Cobban, W. A., and Jeletzky, J. A., 1965, A new scaphite from the Campanian rocks of the western interior of North America: Jour. Paleontology, v. 39, no. 5, p. 794-801, pls. 95-96.

Cobban, W. A., and Reeside, J. B., Jr., 1952, Correlation of the Cretaceous formations of the western interior of the United States: Geol. Soc. Am. Bull. v. 63, p. 1011-1044.

Crandell, D. R., 1950, Revision of Pierre shale of central South Dakota: Am. Assoc. Petroleum Geologists Bull. v. 34, p. 2337-2346.

Curtiss, R. E., 1952, Areal geology of the Isabel quadrangle: South Dakota Geol. Survey Geol. Quad., 1 p.

——— 1954, Areal geology of the Black Horse Butte quadrangle: South Dakota Geol. Survey Geol. Quad., 1 p.

——— 1954, Areal geology of the Firesteel Creek quadrangle: South Dakota Geol. Survey Geol. Quad., 1 p.

Darton, N. H., 1909, Geology and underground waters of South Dakota: U.S. Geol. Survey Water-Supply Paper, no. 227, 156 p.

de Girardin, M. E., 1936, A trip to the Bad Lands in 1849: South Dakota Hist. Rev., v. 1, no. 2, p. 51-78.

Deland, C. E., 1918, Fort Tecumseh and Fort Pierre journal and letter books (abstracted); with notes by Doane Robinson: South Dakota Hist. Coll., v. 9, p. 69-238.

Delevoryas, Theodore, 1964, Two petrified angiosperms from the Upper Cretaceous of South Dakota: Jour. Paleontology, v. 38, no. 3, p. 584-586, pls. 95-96.

Denson, N. M., 1950, The lignite deposits of the Cheyenne River and Standing Rock Indian Reservations, Corson, Dewey, and Ziebach Counties, South Dakota, and Sioux County, North Dakota: U.S. Geol. Survey Circ. 78, 22 p.

Dobbin, C. E., and Reeside, J. B., Jr., 1929, The contact of the Fox Hills and Lance formations: U.S. Geol. Survey Prof. Paper 158-B, p. 9-25.

Dorf, Erling, 1942, Upper Cretaceous floras of the Rocky Mountain region; 2. Flora of the Lance formation at its type locality, Niobrara County, Wyoming: Carnegie Inst. Washington Pub. 508, p. 83-168.

Dunbar, C. O., and Rodgers, John, 1957, Principles of stratigraphy: New York, John Wiley and Sons, 356 p.

Elias, M. K. 1933, Cephalopods of the Pierre formation of Wallace County, Kansas, and adjacent area: Univ. Kansas Sci. Bull., v. 21, no. 9, p. 289-363, pls. 28-42.

Estes, Richard, 1964, Fossil vertebrates from the Late Cretaceous Lance Formation, eastern Wyoming: Univ. California Publ. Geol. Sciences, v. 49, 187 p.

Evans, John, and Shumard, B. F., 1854, Descriptions of new fossil species from the Cretaceous formation of Sage Creek, Nebraska, collected by the North Pacific Railroad Expedition, under Gov. J. J. Stevens: Acad. Nat. Sci. Philadelphia Proc., v. 7, p. 163-164.

—————— 1857, On some new species of fossils from the Cretaceous formation of Nebraska territory: Acad. Sci. St. Louis Trans., v. 1, p. 38-42.

Fisher, Stanley P, 1952, The geology of Emmons County, North Dakota: North Dakota Geol. Survey Bull. 26, p. 1-47.

Goetzmann, W. H., 1959, Army exploration in the American West 1803-1863: New Haven, Yale University Press, 507 p.

Gries, J. P., 1942, Economic possibilities of the Pierre shale: South Dakota Geol. Survey Rept. of Inv. 43, 79 p.

Gunter, Gordon, 1947, Paleoecological import of certain relationships of marine animals to salinity: Jour. Paleontology, v. 21, no. 1, p. 77-79.

Hall, James, and Meek, F. B., 1856, Descriptions of new species of fossils, from the Cretaceous formations of Nebraska, with observations upon *Baculites ovatus* and *B. compressus*, and the progressive development of the septa in baculites, ammonites, and scaphites: Amer. Acad. Arts and Sci. Mem., (new series), v. 5, pt. 2, art. 17, p. 379-411.

Häntzschel, Walter, 1952, Die Lebensspur *Ophiomorpha* Lundgren im Miozan bei Hamburg, ihre weltweit Verbreitung und Synonymie: Hamburg, Geol. Staatsinst., Mitt. H. 21, p. 142-153.

Harris, E., 1845, On the geology of the Upper Missouri: Acad. Nat. Sci. Philadelphia Proc., v. 2, p. 235-240.

Hayden, F. V., 1856, Geologic notes on Nebraska, *in* Warren, G. K., Explorations in the Dacota Country in the year 1855: U.S. 34th Cong., 1st sess., S. Ex. Doc. 76, p. 63-79.

—————— 1857a, Notes explanatory of a map and section illustrating the geological structure of the country bordering on the Missouri River, from the mouth of the Platte River to Fort Benton, in lat. 47° 30'N., long. 110° 30'W.: Acad. Nat. Sci. Philadelphia Proc., v. 9, p. 109-116.

—————— 1857b, Notes on the geology of the Mauvaises Terres of White River, Nebraska: Acad. Nat. Sci. Philadelphia Proc., v. 9, p. 151-158.

—————— 1858, Explanations of a second edition of a geological map of Nebraska and Kansas, based upon information obtained in an expedition to the Black Hills, under the command of Lieut. G. K. Warren, Top. Engr. U.S.A.: Acad. Nat. Sci. Philadelphia Proc., v. 9, p. 139-158.

—————— 1862, On the geology and natural history of the Upper Missouri: Am. Philos. Soc. Trans., v. 12, art. 1, p. 1-218.

—————— 1869, Geological report of the exploration of the Yellowstone and Missouri Rivers; under the direction of Capt. W. F. Raynolds, Corps of Engineers, 1859-60: U S. 40th Cong., 2nd sess., S. Ex. Doc. 77, 174 p.

Howarth, M. K., 1965, Cretaceous ammonites and nautiloids from Angola: British Mus. Nat. Hist. Bull., v. 10, no. 10.

Jeletzky, J. A. 1960, Youngest marine rocks in western interior of North America and the age of the *Triceratops*-beds; with remarks on comparable dinosaur-bearing beds outside North America: Internat. Geol. Cong., 21st, Copenhagen, Rept., pt. 5, p. 25-40.

—————— 1962, The allegedly Danian dinosaur-bearing rocks of the globe and the problem of the Mesozoic-Cenozoic boundary: Jour. Paleontology, v. 36, no. 5, p. 1005-1018, pl. 141.

Jeletzky, J. A., and Clemens, W. A., 1965, Comments on Cretaceous Eutheria, Lance *Scaphites*, and *Inoceramus?* ex. gr. *Tegulatus*: Jour. Paleontology, v. 39, no. 5, p. 952-959.

Laird, W. M., and Mitchell, R. H., 1942, The geology of the southern part of Morton County, North Dakota: North Dakota Geol. Survey Bull. 14, p. 1-42.

Leidy, Joseph, 1856, Notices of extinct Vertebrata discovered by Dr. F. V. Hayden, during the expedition to the Sioux country under the command of Lieut. G. K. Warren: Acad. Nat. Sci. Philadelphia Proc., v. 8, p. 311-312.

Lovering, T. A., and others, 1932, Fox Hills formation, northeastern Colorado: Am. Assoc. Petroleum Geologists Bull., v. 16, p. 702-703.

Manz, O. E., 1962, Investigation of pozzolanic properties of the Cretaceous volcanic ash deposit near Linton, North Dakota: North Dakota Geol. Survey Rept. Inv. 38, 42 p.

Meek, F. B., 1876, A report on the invertebrate Cretaceous and Tertiary fossils of the upper Missouri country: U.S. Geol. Survey Terr., v. 9, 629 p.

Meek, F. B., and Hayden, F. V., 1856a, Descriptions of new species of Gasteropoda from the Cretaceous formations of Nebraska Territory: Acad. Nat. Sci. Philadelphia Proc., v. 8, p. 63-69.

—————— 1856b, Descriptions of new species of Gasteropoda and Cephalopoda from the Cretaceous formations of Nebraska Territory: Acad. Nat. Sci. Philadelphia Proc., v. 8, p. 70-72.

—————— 1856c, Descriptions of twenty-eight new species of Acephala and one gasteropod, from the Cretaceous formations of Nebraska Territory: Acad. Nat. Sci. Philadelphia Proc., v. 8, p. 81-87.

—————— 1856d, Descriptions of new species of Acephala and Gasteropoda, from the Tertiary formations of Nebraska Territory, with some general remarks on the geology of the country about the sources of the Missouri River: Acad. Nat. Sci. Philadelphia Proc., v. 8, p. 111-126.

———— 1856e, Descriptions of new fossil species of Mollusca collected by Dr. F. V. Hayden, in Nebraska Territory; together with a complete catalogue of all the remains of Invertebrata hitherto described and identified from the Cretaceous and Tertiary formations of that region: Acad. Nat. Sci. Philadelphia Proc., v. 8, p. 265-286.

———— 1857, Descriptions of new species and genera of fossils, collected by Dr. F. V. Hayden in Nebraska Territory, under the direction of Lieut. G. K. Warren, U.S. Topographical Engineer; with some remarks on the Tertiary and Cretaceous formations of the north-west, and the parallelism of the latter with those of other portions of the United States and Territories: Acad. Nat. Sci. Philadelphia Proc., v. 9, p. 117-148.

———— 1861, Descriptions of new Lower Silurian, (Primordial), Jurassic, Cretaceous, and Tertiary fossils, collected in Nebraska, by the exploring expedition under the command of Capt. Wm. F. Raynolds, U.S. Top. Engrs.; with some remarks on the rocks from which they were obtained: Acad. Nat. Sci. Philadelphia Proc., v. 13, p. 415-447.

Moore, D. G., and Scruton, P. C., 1957, Minor internal structures of some recent unconsolidated sediments: Am. Assoc. Petroleum Geologists Bull., v. 41, no. 12, p. 2723-2751.

Moore, R. C., 1957, Modern methods of paleoecology: Am. Assoc. Petroleum Geologists Bull., v. 41, p. 1775-1801.

Morgan, R. E., and Petsch, B. C., 1945, A geological survey in Dewey and Corson Counties, South Dakota: South Dakota Geol. Survey Rept. Inv. 49, p. 1-45.

Nicollet, J. N., 1841, The geology of the region on the Upper Mississippi, and the Cretaceous formation of the upper Missouri: Am. Jour. Sci., v. 41, p. 180-182.

———— 1843, On the Cretaceous formation of the Missouri River: Am. Jour. Sci., v. 45, no. 1, p. 153-157.

Nota, D. J. G., 1958, Sediments of the Western Guiana Shelf, in Reports of the Orinoco Shelf Expedition, v. 2: Utrecht, Meded. Landbouw. te Wageningen, v. 58, no. 2, p. 1-98.

Nowak, Jean, 1911, Untersuchungen über die Cephalopoden der oberen Kreide in Polen. 2. Teil: Die Skaphiten: Acad. Sci. Cracovie Bull. Internat., ser. B, n. 7, p. 547-589.

Owen, D. D., 1852, Report of a geological survey of Wisconsin, Iowa, and Minnesota; and incidentally of a portion of Nebraska Territory: Philadelphia, Lippincott, Grambo, 638 p.

Petsch, B. C., (compiler), 1953, Geologic map of South Dakota: South Dakota Geol. Survey.

Pettyjohn, W. A., 1961, Geology of the Glencross quadrangle: South Dakota Geol. Survey Geol. Quad., 1 p.

———— 1967, New members of the Upper Cretaceous Fox Hills Formation in South Dakota, representing delta deposits: Am. Assoc. Petroleum Geologists Bull., v. 51, no. 7, p. 1361-1367.

Prout, H. A., 1847, Description of a fossil maxillary bone of *Palaeotherium* from near White River: Am. Jour. Sci., 2nd ser., v. 3, no. 8, 248-250.

Robinson, C. S., Mapel, W. J., and Cobban, W. A., 1959, Pierre shale along western and northern flanks of Black Hills, Wyoming and Montana: Am. Assoc. Petroleum Geologists Bull. v. 43, p. 101-123.

Russell, W. L., 1925a, Well log in northern Ziebach County: South Dakota Geol. and Nat. Hist. Survey, Circ. 18, p. 1-14.

———— 1925b, The possibilities of oil in western Ziebach County: South Dakota Geol. and Nat. Hist. Survey, Circ. 20.

———— 1926, The possibilities of oil in western Corson County: South Dakota Geol. and Nat. Hist. Survey, Circ. 27, p. 1-18.

Schlüter, Clemens, 1871-76, Die Cephalopoden der oberen deutschen Kreide: Palaeontographica, v. 21 (1871-72), p. 1-120, pls. 1-35; v. 24, (1876), p. 121-264, pls. 36-55.

Searight, W. V., 1931, The Isabel-Firesteel coal area: South Dakota Geol. Survey Rept. Inv. 10, p. 1-35.

———— 1934, The Stoneville coal area: South Dakota Geol. Survey Rept. Inv. 22, p. 1-20.

———— 1937, Lithologic stratigraphy of the Pierre formation of the Missouri Valley in South Dakota: South Dakota Geol. Survey Rept. Inv. 27, p. 1-63.

Shaw, A. B., 1964, Time in stratigraphy: New York, McGraw-Hill, 350 p.

Shepard, F. P., 1960, Mississippi delta: marginal environments, sediments, and growth, in Shepard, F. P., Phleger, F. B., and van Andel, T. H., eds., Recent sediments, northwest Gulf of Mexico: Tulsa, Okla., Am. Assoc. Petroleum Geologists, p. 56-81.

Stanton, T. W., 1910, Fox Hills sandstone and Lance formation ("Ceratops beds") in South Dakota, North Dakota and eastern Wyoming: Am. Jour. Sci., v. 30, p. 172-188.

———— 1917, A Cretaceous volcanic ash bed on the Great Plains in North Dakota: Washington Acad. Sci. Jour., v. 7, p. 80-81.

———— 1920, The fauna of the Cannonball marine member of the Lance formation: U.S. Geol. Survey Prof. Paper 128-A, p. 1-60.

Stevenson, R. E., 1956, Areal geology of the Bullhead quadrangle: South Dakota Geol. Survey Geol. Quad., 1 p.

———— 1957, Geology of the McIntosh quadrangle: South Dakota Geol. Survey Geol. Quad., 1 p.

———— 1959, Geology of the Miscol quadrangle: South Dakota Geol. Survey Geol. Quad., 1 p.

———— 1960a, Geology of the Timber Lake quadrangle: South Dakota Geol. Survey Geol. Quad., 1 p.

———— 1960b, Geology of the Little Eagle quadrangle: South Dakota Geol. Survey Geol. Quad., 1 p.

Teichert, Curt, 1958, Concepts of facies: Am. Assoc. Petroleum Geologists Bull., v. 42, p. 2718-2744.

Thom, W. T. Jr., and Dobbin, C. E., 1924, Stratigraphy of Cretaceous-Eocene transition beds in eastern Montana and the Dakotas: Geol. Soc. America Bull., v. 35, p. 481-506.

Thorson, Gunnar, 1957, Bottom communities (sublittoral or shallow shelf), in Treatise on marine ecology and paleoecology: Geol. Soc. America Mem. 67, v. 1, p. 461-534.

Todd, J. E., 1910, Preliminary report on the geology of the northwest-central portion of South Dakota, in Report of the State Geologist for 1908: South Dakota Geol. Survey Bull. 4, p. 13-76, 193-207.

Tychsen, P. C., and Vorhis, R. C., 1955, Reconnaissance of geology and ground water in the lower Grand River valley, South Dakota: U.S. Geol. Survey Water-Supply Paper, no. 1298, 33 p.

van Straaten, L. M. J. U., 1959, Minor structures of some recent littoral and neritic sediments: Geologie en Mijnbouw (new ser.), v. 21, p. 197-216.

Waage, K. M., 1961, The Fox Hills Formation in its type area, central South Dakota, in Symposium on Late Cretaceous rocks: Wyoming Geol. Assoc. 16th Ann. Field Conf. Guidebook, p. 229-240.

———— 1964, Origin of repeated fossiliferous concretion layers in the Fox Hills Formation: Kansas Geol. Survey Bull. 169, p. 541-563. [1966].

———— 1965, The Late Cretaceous coleoid cephalopod Actinosepia canadensis Whiteaves: Peabody Mus. Nat. Hist. (Yale Univ.) Postilla, no. 94, 33 p.

———— 1967, Cretaceous transitional environments and faunas in central South Dakota; in Symposium on paleoenvironments of the Cretaceous seaway in the western interior: Preprints of Papers, Colo. School of Mines, p. 237-266.

Ward, F., and Wilson, R. A., 1922, Possibilities of oil in western Dewey County: South Dakota Geol. and Nat. Hist. Survey, Circ. 9.

Warren, G. K., 1856, Explorations in the Dacota Country in the year 1855: U.S. 34th Cong., 1st sess., S. Ex. Doc. 76, 79 p.

———— 1859, Preliminary report on explorations in Nebraska and Dakota in the years 1855-'56-'57: U.S. 35th Cong., 2nd sess., H. Ex. Doc. 2 (reprinted Washington, D.C., U.S. Govt. Printing Office, 1875,) 125 p.

Weimer, R. J., and Hoyt, J. H., 1964, Burrows of Callianassa major Say, geologic indicators of littoral and shallow neritic environments: Jour. Paleontology, v. 38, no. 4, p. 761-767.

Weller, J. M., 1960, Stratigraphic principles and practice: New York, Harper Brothers, 678 p.

Wells, J. W., 1947, Provisional paleoecological analysis of the Devonian rocks of the Columbus region: Ohio Jour. Sci., v. 47, p. 119-126.

Wheeler, H. E., and Mallory, V. S., 1956, Factors in lithostratigraphy: Am. Assoc. Petroleum Geologists Bull., v. 40, p. 2711-2723.

Willis, Bailey, 1885, The lignites of the Great Sioux Reservation; a report on the region between the Grand and Moreau Rivers, Dakota: U.S. Geol. Survey Bull. 21, 16 p.

Wilmarth, M. G., 1938, Lexicon of geologic names of the United States (including Alaska): U.S. Geol. Survey Bull. 896, 2396 p.

Wilson, R. A., 1922, The possibilities of oil in northern Dewey Country: South Dakota Geol. and Nat. Hist. Survey Circ. 10, p. 1-8.

———— 1925. The Ragged Butte structure: South Dakota Geol. and Nat. Hist. Survey Circ. 24, 7 p.

Wilson, R. A., and Ward, Freeman, 1923, The possibilities of oil in northern Ziebach County: South Dakota Geol. and Nat. Hist. Survey Circ. 13, 11 p.

Young, Keith, 1960, The Cretaceous ammonite successions of the Gulf Coast of the United States: Internat. Geol. Cong., 21st, Copenhagen, Rept., pt. 21, p. 251-260.

Ziegler, Bernhard, 1963, Ammoniten als Faziesfossilien: Paläont. Zeitschr., v. 37, p. 96-102.

UNPUBLISHED SOURCES

Meek, F. B., 1853, Journal of Nebraska Expedition: Field Book 3, Fielding B. Meek Papers, Smithsonian Institution.

Mello, J. F., 1962, Stratigraphy and micropaleontology of upper Pierre Shale in north-central South Dakota: Ph.D. dissertation, Yale University.

Raynolds, W. F., 1859-60, Field journals, Yellowstone Expedition: William F. Raynolds Papers, Yale University, Western Americana Collection.

Speden, I. G., 1965, Paleozoology of Lamellibranchia from the type area of Fox Hills Formation (Upper Cretaceous, Maestrictian), South Dakota: Ph.D. dissertation, Yale University.

PLATE 1
Trail City Member; Little Eagle lithofacies

A. Lower part of type Little Eagle lithofacies at Locality 50. Upper man stands on light streak of jarositic silt that marks base of Fox Hills Formation; darker shale of Pierre, obscured by wash on slope, shows in lower right.

B. Color contrast between Elk Butte Member of Pierre Shale (left) and Trail City Member of Fox Hills (right) is evident along fault in road cut north of Little Eagle. Concretions of Lower *nicoleti* Assemblage Zone are evident on the Trail City exposure.

C. Gulley at Locality 30; part of exposures constituting principal reference section of Trail City Member. Concretion layer of *Protocardia — Oxytoma* Assemblage Zone (PZ) and barren A (BA) layer are evident. Man stands at top of light-colored sandy zone that contains the medial jarositic silt.

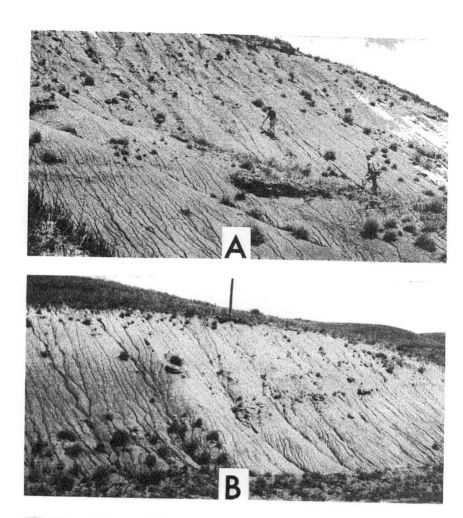

light
erre,

Mem-
Eagle.
sure.
Trail
and
that

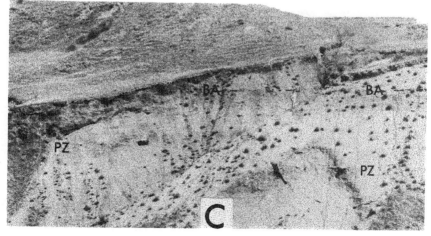

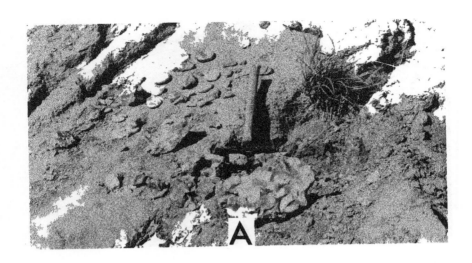

PLATE 2
Fossils from the Trail City Member, Little Eagle lithofacies

A. Typical Lower *nicoletti* concretion dug from exposure shown in Pl. 1, fig. B.
B. Close-up of fragment of a Lower *nicoletti* concretion showing specimens of *Scaphites (Hoploscaphites) nicolleti.*
C. Small concretion from marginal area of Lower *nicolleti* Assemblage Zone containing single specimen.

PLATE 3
Fossils from the Trail City Member, Little Eagle lithofacies
A. Small *Limopsis* and large *Gervillia* concretions found weathered out of beds in *Limopsis — Gervillia* Assemblage Zone.
B. Fragment of a *Protocardia — Oxytoma* concretion with abundant *Oxytoma nebrascana* and scattered *Protocardia subquadata*.
C. *Protocardia — Oxytoma* concretion partially dug from the outcrop.

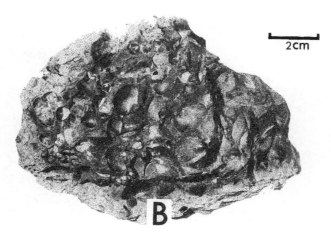

2cm

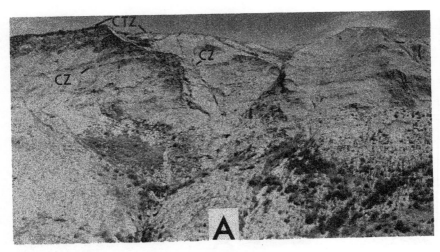

PLATE 4

Trail City Member, Irish Creek lithofacies; and Timber Lake Member.

A. Bluffs of Irish Creek lithofacies on north side of Moreau River at Locality 210 in Ziebach County. Steep upper part includes beds equivalent to Timber Lake Member, which terminates about 1 mile east of this locality. Principal concretion layer (CZ) of *Cucullaea* Assemblage Zone lies approximately 30 feet below concretions in *Cymbophora — Tellina* Assemblage Zone (CTZ) which cap the bluff.

B. Timber Lake Member at Locality 275 on Standing Cloud Creek. Note change in bedding from thick and massive below concretions at CT (level of *Cymbophora — Tellina* Assemblage Zone) to more tabular *Tancredia — Ophiomorpha*-bearing beds above. X marks contact with Iron Lightning Member. (see section 12, p. 114).

C. Upper part of Rock Creek lithofacies of Timber Lake Member at its type exposure (Loc. 194) near Bullhead. Gradation into more typical Timber Lake sand begins at small concretionary lens about 8 feet below man.

A. *Ophiomorpha* — bearing sand of Timber Lake Member at Locality 48 (see section 11 unit 7, p. 113).

B. Planar upper contact of Timber Lake Member with Bullhead lithofacies of the Iron Lightning Member along Firesteel Creek, about 2 miles south of Locality 275.

C. Indurated sandstone lens at top of Timber Lake Member south of Grand River near Locality 132. Note planar contact with Iron Lightning Member.

ection 11

the Iron

er near

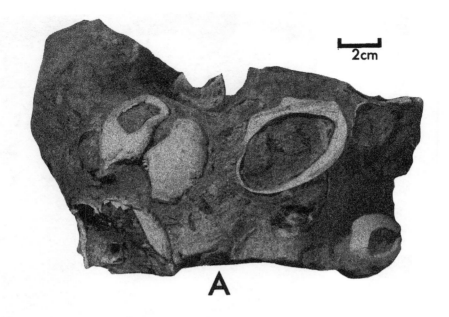

A

B

PLATE 7
Fossils from the Timber Lake Member

A. Concretion with complete specimen of *Sphenodiscus lenticularis* from Locality 35.

B. Concretion with large *Discoscaphites nebrascensis* from Locality 35.

C. *Pteria linquaeformis*; left, in *Pteria-Ostrea* association from level of *Cymbophora-Tellina* Assemblage Zone along Firesteel Creek; right, clustered in probable growth position around axis — from upper right to lower left; elongate object (plant?) to which they were attached is no longer preserved. From concretion in *Cucullaea* Assemblage Zone.

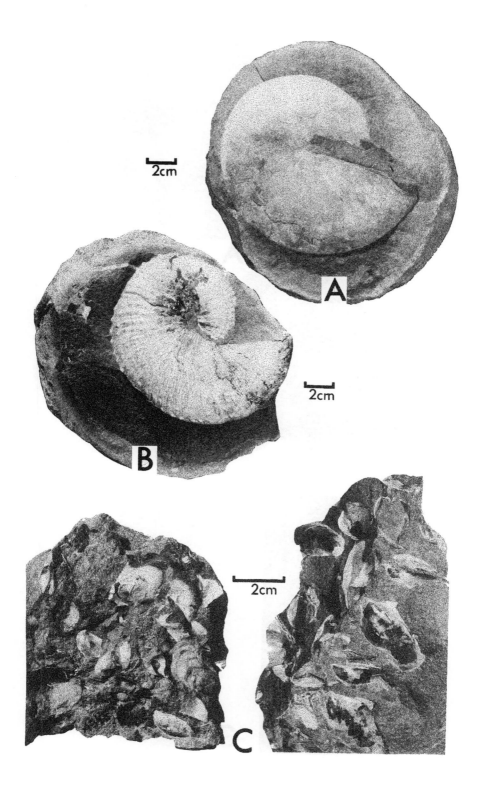

A

2cm

B

2cm

C

2cm

2cm

A. *Cymbophora* (center) — *Tellina* (upper right) association with plant fragments; from concretion in Irish Creek lithofacies at level of *Cucullaea* Assemblage Zone along Firesteel Creek. (see section 12, unit 1, p. 114).

B. *Tancredia americana* from the *Tancredia* — *Ophiomorpha* biofacies of Timber Lake Member. Paired valves, in probable living position, show posterior gape for large siphons.

C. *Ophiomorpha* tube (left), in natural longitudinal section, shows crustacean appendage at bottom. Right, enlargement of crustacean fragment identified (by H. B. Roberts of the U.S. National Museum) as pincer claw, wrist and distal part of arm of burrowing shrimp, *Callianassa* sp.

PLATE 9
Iron Lightning Member
A. Bullhead lithofacies of Iron Lightning in contact (at hammer handle) with sand of Timber Lake Member, Locality 48; shown are parts of units 8 and 9, section 11 (p. 113).
B. Typical banded aspect of Bullhead lithofacies; upper man at zone of contorted bedding at base of overlying Colgate lithofacies.
C. Biogenic structures in the Bullhead lithofacies. Left; canoe-shaped trough with concentric, laminated, filling. Note *Dentalium* fragments (D.) Right; fecal pellets revealed on fresh-cut surface in plane of bedding.

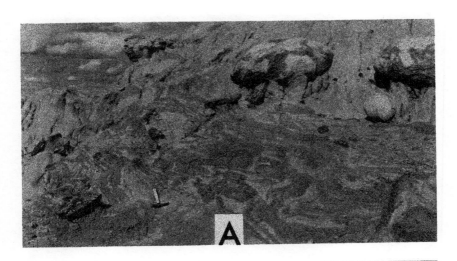

Iron Lightning Member

A. Local area of contorted beds in Bullhead lithofacies below a sand body of the Colgate lithofacies at Locality 48.

B. Part of the Iron Lightning badlands, showing local dominance of Bullhead lithofacies in Iron Lightning Member. Colgate lithofacies weakly present near top of member (U) and a thin second zone of Colgate is indicated by the white-weathering concretionary sandstone masses extending left from L.

C. Close-up of exposure of sandy beds at the lower (L) Colgate horizon shown in figure B above. This widespread, thin sandy unit in the Bullhead lithofacies is a useful key bed in the western part of the type area.

PLATE 11
Iron Lightning Member and the Fox Hills-Hell Creek contact

A. Trough crossbedding in Colgate lithofacies near Locality 48 along Hump Creek.

B. Colgate lithofacies (remnant in lower left) cut out by channel filled with Bullhead-like lithology just below lignitic clay marking contact (C) of Fox Hills with Hell Creek.

C. Fox Hills-Hell Creek contact (C) at Locality 48, bluffs south of Hump Creek. The locally thick Colgate lithofacies has characteristic channel-fill structure and large concretions. Note other beds of Colgate lithofacies above basal lignitic clay of Hell Creek.

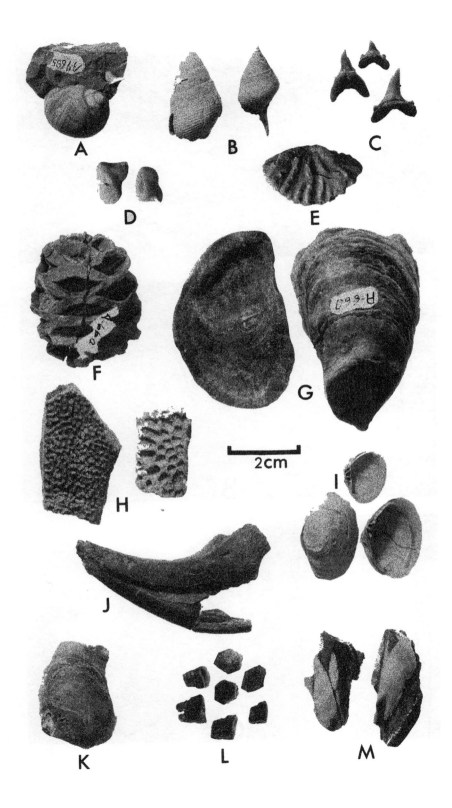

A

B

C

D

E

F

G

2cm

H

I

J

K

L

M

Fossils from the Iron Lightning Member. A thru E from Bullhead lithofacies, F thru M from Colgate lithofacies. All from the Iron Lightning badlands, Locality 74.

A. *Lunatia concinna*
B. *Piestochilus scarboroughi*
C. Shark teeth
D. *Vorhisia vulpes*, fish otoliths
E. Fragment of *Discoscaphites*
F. Conifer cone, probably *Sequoia dakotensis*
G. *Crassostrea subtrigonalis*
H. Turtle and crocodile scutes
I. *Corbicula subelliptica*
J. Claw of carnivorous dinosaur
K. *Anomia gryphorhyncha*
L. *Myledaphis bipartitus*, tooth elements of ray-like fish
M. *Mylognathus priscus*, dental plates of mollusk-eating fish

INDEX

(**Boldface** numbers indicate illustrations; plate and figure numbers are preceded by **pl.** and **fig.**).

173